JUSTIC
MALCOLM

HUMAN SACRIFICE UNDER U.S. BIOTERRORISM

JOANY CHOU, Ph.D.

Published @ 2014 Joany Chou, Ph.D.
Printed by CreateSpace, An Amazon Company
CreateSpace, Charleston, SC

Library of Congress Cataloging-in-Publication Data

Cover Photo by Casia Holmgren in June 2009
Cover Design by Justin Carder
ISBN-13: 978-0-9905446-0-9
ISBN-10: 0990544605
Printed by CreateSpace, An Amazon Company
CreateSpace, Charleston, SC
eStore: www.CreateSpace.com/4906966
Available on Kindle and other Retail Outlets

DEDICATION

To John and Dolores

To John and Dolores for the gift of their son Malcolm to us, to the family, to Beckwith, to his students, to friends and colleagues at the University, and to countless other scientists all over the world who studied infectious diseases and molecular biology. And to those who suffered human injustice and human rights exploitation from the development of U.S. bioterrorism.

Contents

Prologue

SEPTEMBER 13, 2009

The ambulance raced into the Chicago Hospital parking lot like it was the end of the world and parked outside emergency. The two paramedics hurried around back and lifted the man onto a transport dolly.

The man, Malcolm Casadaban, was rushed into the ER and examined thoroughly by the doctors. He was experiencing fever, a racking cough, and shortness of breath. The paramedics had recorded an oxygen saturation level of 92 percent, so doctors quickly administered oxygen through a mask.

Malcolm's mind frantically tried to figure out what had happened to him, what could be causing the total collapse of his body. But his head hurt so badly, like a knife was piercing it, that he could not think. Everything was cloudy. Whatever it was, he knew it was bad. He looked for his family, his daughters, or any other familiar faces, people who could offer comfort and aid. He was alone in that dark night on a stretcher surrounded by white coats of doctors and nurses who rushed to attend him in the University Emergency Room in Bernard Mitchell Hospital. This was September 13, 2009.

The doctors scurried about and struggled to determine the prognosis of the illness. He just lay there in his bed as the pain increased and shot through his head and body like some unholy thing. A few days earlier, he had told laboratory colleagues that he was coming down with flu-like symptoms, but this couldn't be a simple flu. Initially, he appeared alert and was able to speak, answering questions. But he struggled to breathe, gasping for air, and even with an oxygen mask each breath came with ever-increasing difficulty. Eventually, he found himself unable to complete even one sentence. His temperature climbed to 100 degrees F and his heart beat faster, 106 beats per minute. He was retaining a great deal of water. Cyanic color appeared on his nail beds.

The doctors, feeling they had an idea about what was happening to their sick patient, Malcolm was treated for congestive heart failure, water in the lung, and liver failure and you name it, and gradually all bodily organs began to shut down. Nothing worked; gradually all his organs began to shut down. They gave him morphine, Ativan, and Propofol to calm him down and ease his pain. By this time, doctors had discovered that Malcolm's blood was turbid, badly contaminated with bacteria, a condition so deadly that Malcolm was clearly on a no-return road. The doctors started a stringent antibiotic administration at noon, roughly nine hours after Malcolm had been admitted. But with this prognosis, the doctors knew that Malcolm was lingering in his last few hours on earth. They increased the dosage of morphine to keep him focused and comfortable. By this time, the doctors had found abdominal distention, peripheral cyanosis, and trace pedal edema. While confusing, at least these symptoms were something the doctors could relate to and treat; or so they thought.

Malcolm realized by now what has finally taken place in his body, but he knew no treatment would avail at this late stage of his blood infection. He tried to tell his attending nurses what he thought was happening to him, but lack of air prevented him from completing even a sentence. He looked for his family, anyone who could understand him, but no one was there. As hope dwindled, as his body began to give up the struggle, a wicked realization pushed to the forefront of his mind. Perhaps this wasn't something natural after all. Perhaps it was plague that was driving his body to the end. Perhaps this infection was some kind of twisted revenge against him, against his family.

Now Malcolm's stomach was horribly distended. His lungs were filling with water that pushed out the air he so desperately needed. His whole body began to turn blue, signaling lack of oxygen. He was sweating profusely. His face was desperate and he sobbed. He looked for his family, but no one was there.

At about 5 p.m., one of the doctors handed Malcolm a form to sign giving them permission for intubation, a last ditch effort to reduce fluid in his lungs to assist his breathing. Malcolm was in no condition to refuse anything at that point. With some assistance, he shakily wrote his name and uttered his final words, "Do what you have to do."

A malicious, silent killer lurked just beneath the surface of this man's ravaged body. On September 13, 2009, at 5:08 p.m., twelve hours after he had been rushed to the hospital, Malcolm gave his last breath and fell eternally silence. He was pronounced dead. No one knew why.

1

THE IMPOSSIBLE HAPPENS

There are moments in a person's life that are so monumental, so devastatingly significant that they stand out like beacons among every other memory a person has. They linger as nightmares and keep you awake at night. They enter your life like thieves and plunder and ravage every part of your heart and your soul. This is my story and it is not for the faint of heart. Every word is true. Every word is spoken from Malcolm's gentle spirit inside of me.

It began like any ordinary day, that Sunday afternoon in September 2009, when the phone call came. I remember it like it was yesterday because the pain is still excruciating and the agony in my daughter's voice was burning.

I was visiting my elderly mother at a nursing home near my home in California. Brooke often called me on Sunday afternoons to chat and check up on us. She was in Las Vegas for the weekend.

"Oh, Mom!" she shrieked, sobbing uncontrollably.

"What is it Brooke? What's wrong?" I asked, immediately alarmed. *Had someone been in a car accident? Had someone had a heart attack?*

She could hardly speak. Her tone was horrifying, and her words slurred. My heart sank as I struggled to understand any part of what she was trying to say, but I couldn't.

"Mommmmm," she mumbled through the tears. "Oh, Mommmmm."

"Brooke!" I yelled. "I don't understand what you're telling me! Calm down and tell me what's going on."

"I juss … Mom … Da…" She slurred and garbled more words. Still, I couldn't understand what she was saying.

Then, I heard the words, "Dad . . . Dad . . . dead." Those words were like a piercing knife in my heart. For a moment I couldn't speak, still hoping what I heard was wrong.

"Malcolm died?" I finally forced out, hoping for a denial, hoping I had misheard.

Brooke whimpered, "Yes."

That moment changed my life forever.

September 13, 2009, would mark the day that Malcolm Casadaban, a brilliant molecular geneticist and professor at the University of Chicago, my former husband of twenty-seven years, the father of my two children and my best friend, had died at the age of sixty at the university's Bernard Mitchell Hospital. That day would mark the beginning of a vast change in my life, altering it in ways I could never have imagined.

Even though my heart was pounding, I kept my voice calm and said, "Brooke, what is going on? I can't understand you, Brooke, speak a little slower."

Brooke took a deep breath, and then said, "Mom, I received two calls from the Chicago ER doctors. On the first call, they said they were trying to resuscitate Dad. Then, they called me back again—just a minute ago—and said that Dad had just passed away at 7:08 p.m., Chicago time." That was 5:08 p.m. in California.

"7:08 p.m., Sunday evening, September 13, 2009." These words kept repeating in my mind, as if by holding on to that moment, I could hold on to Malcolm before he crossed the threshold from life to death. My tears started to flow. Nothing made sense.

"Come home, come home quickly, Brooke. We need to be on the next plane to Chicago. Let's go together," I told her. I knew we had to go see Malcolm, that it was the only thing to do, and I knew we needed to go together. We needed each other so desperately in this moment. "Hurry home ... hurry home." I found myself repeating these words because I had no words of comfort to offer her right then.

At that moment, I realized how fragile our lives were, lives that had been broken by the harsh reality of a single phone call. Malcolm was gone, and my life and the lives of our daughters were forever changed.

After we hung up, I sat there trying to calm myself and figure out what could have happened to strike down a sixty-year-old man who had been in good health. A range of scenarios ran through my mind, from the mundane—*Was Malcolm in a car accident?*—to the lurid—*Was he accidentally killed in a South Side Chicago gang shooting?*

I had no idea what had happened, and my mind struggled to make some sense of it. I started to pray for Malcolm, hoping that he had finally found peace because; it had seemed to me, that something had been upsetting him in recent months.

I remembered the pledge Malcolm had asked me to make just two months ago: in the event that one of us died, the other would assume the role of caretaker of our two children. Malcolm had always loved and adored his two daughters above anything else, but I remember thinking his words were odd, even ominous. After all, he was only sixty years old. He had yet to give his daughters away at their weddings and bestow hugs on his grandchildren. He had yet to see Leigh become a surgeon. He had had so much to look forward to and should have had time to see and do all of these things. But his life had been cut short.

My thoughts ranged further. I wondered if Malcolm knew he was dying and if he had, when. What did he tell the doctors in the ER before his death? Did he look for his family while he lay dying?

I shook myself. I would find no answers sitting in my mother's room. This was a family crisis, and I needed to pull myself together to figure out precisely how I was going to spend the next twenty-four-hours. I reminded myself to be strong—strong enough to handle all the grim details, strong enough to help my daughters through the ordeal, strong enough to take the journey back to Chicago, to a city that I had left behind when Malcolm and I were divorced, and strong enough to bring him home.

I said good-bye to my mother and drove home to start making painful phone calls to all the relatives. I called Malcolm's mother, Dolores. I called his brothers and sisters who lived in New Orleans to tell them of the shocking news that Malcolm had passed away.

I had no idea, yet, just how strong I would have to be to discover what had happened to Malcolm, no idea that the cause and circumstances of his death lay buried in deep shadows, no idea that there were those who did not want the truth to come to light.

2

"HE WAS IN ANOTHER WORLD IN SOME WAYS, BUT WHAT A WORLD HE CREATED"

Brooke and I boarded our plane at midnight at the LAX Airport. We clung to each other and sobbed quietly, saying very little to each other. The sorrow in our hearts was too much to bear. *I had only talked to him just weeks ago! How could this be . . .?*

When the plane began to take off, I sank in my seat and closed my tired, puffy eyes. Thoughts and memories of Malcolm flashed through my mind, all ages, in so many places, at so many points in our lives. I saw him in his mid-twenties, a young man with sparkling eyes filled with confidence and determination, his hair curly, and his build as thin as a rail. He was talking to another scientist and was excited about his theories.

It was 1975 in Dr. Stanley Cohen's laboratory at Stanford Medical School that I first saw Malcolm. I had just graduated from University of California at Berkeley the previous year and was fascinated by the development of recombinant DNA technology in Cohen's laboratory. It was the era of cloning, cloning in its physical sense of cutting and joining of DNA segments *in vitro,* along with the creation of the first chimeric plasmid, pSC101, in the spring of 1973. The next year the first Boyer and Cohen patent was filed by the University of California at San Francisco and Stanford University; the first Asilomar Conference on recombinant DNA was held in February 1975. I was captivated by the new art. It was indeed the dawning age of molecular biology when cloning revolutionized our way of thinking and doing science.

When Malcolm walked into Cohen's lab that day in 1976, he was not exactly the princely figure that I had dreamed of, to say the least. I saw a strange fellow who resembled a hippie more than a brilliant scientist. He had long, unkempt curly hair pulled back into a small ponytail and a long twisted beard that almost completely covered up his thin lips. He was dressed like a typical impoverished Harvard intellect. Why not? That was precisely the image he relished. Yet underneath all that hair and beard was a rather handsome face to behold. His eyes sparkled with confidence and determination. He was twenty-six when I met him.

Malcolm working on bench in Stanford Laboratory

Before Malcolm had even set foot in Stanford, there had been whispers that the soon-to-arrive postdoctoral was a man of genius. By the size of his reputation, I assumed that this new postdoc would be an older man. At the age of twenty-six, Malcolm had already made a name for himself as a brilliant young scientist, having earned renown while as a graduate student in the laboratory of Dr. Jonathan Beckwith at Harvard Medical School

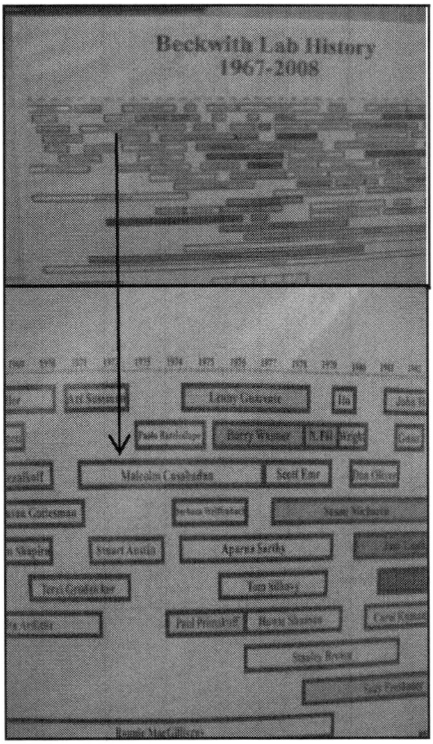

Time capsule of students and researchers working in Beckwith's laboratory in 1967—2008.

As I got to know him, I discovered that Malcolm was a gentle soul who loved science above anything in this world. He was soft-spoken—so much so that you had to lean close to hear what he had to say. Underneath the air of strong determination lay a shy, awkward, and somewhat nerdish man (in that way, a typical MIT graduate). Sometimes, in the middle of an intense conversation with another scientist, he would abruptly break off, looking far into the sky, then seconds later, he would return with a new dialogue, as if God had just flashed him with another brilliant idea. Malcolm was brilliant in every sense of the word.

Gene Fusion in Early Development

In 1971, Malcolm, a young and daring graduate student, entered the laboratory of Dr. Jonathan Beckwith at Harvard Medical School. He had just graduated from neighboring Massachusetts Institute of Technology (MIT) that summer, and like many other science graduate students, he was wondering how a gene turned itself on and off in the presence of biological substrates such as maltose in a maltose operon. Malcolm, enthused with the basic biology, was in the right place, in the midst of mainstream science at Harvard Medical School.

Like other science students in his time, Malcolm resorted to complicated phage genetics to take on his study. Bacteriophage had the ability to disrupt, juxtaposed and rearranged genes like what nature would do at that time.

As a fresh graduate student in Beckwith's laboratory, Malcolm enjoyed a rare freedom to wander about in the different realms of science to find a system that he could concentrate on. Soon he found out it was not feasible to do his work in the ways then in use. So he hit upon an idea, a novel technology that would change the way the molecular biologists did science. It was a tool to study gene expression and gene regulation and to resolve the mysteries of genetic circuits that allowed the organism to grow and adapt to the environmental stimuli.

Through much trial and error, in 1975, Malcolm developed a gene fusion technology in Jon Beckwith's laboratory that was revolutionary; changing the way molecular biologists did science. For him, the basic concept was simple: it was a method to join indicator-reporter gene such as lacZ that expresses ß-galactosidase, for example, to any promoter in a biologically defined system. Bacteria would then appear in color variation by the regulated lacZ expression in accordance with promoter specificity and strength. Three of his classic papers in 1975 and 1976 set the original concept of gene fusion in history.

Casadaban, M. J. 1975. Fusion of the *Escherichia coli lac* genes to the *ara* promoter: a general technique using bacteriophage Mu-1 insertions. Proc. Natl. Acad. Sci. U. S. A. **72:**809–813.
Casadaban, M. J. 1976. Regulation of the regulatory gene for the arabinose pathway, *araC*. J. Mol. Biol. **104:**557–566.
Casadaban, M. J. 1976. Transposition and fusion of the *lac* genes to selected promoters in *Escherichia coli* using bacteriophage lambda and Mu. J. Mol. Biol. **104:**541–555.

Figure 2A: Original concept of gene fusion by Malcolm Casadaban

The schematic diagram of gene fusion can be found online at http://en.wikipedia.org/wiki/Reporter_gene (Figure 2B).

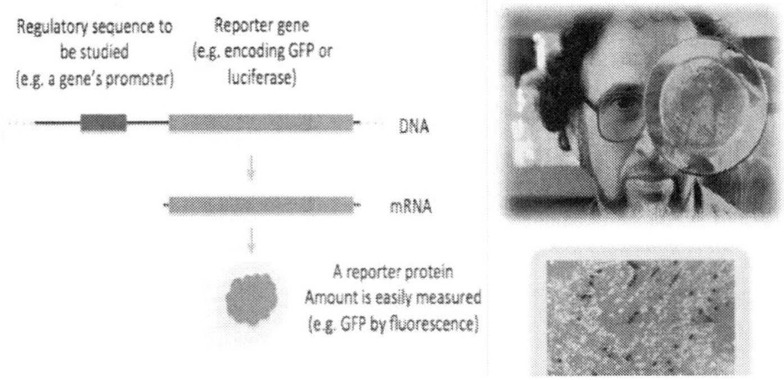

Figure 2B: Diagram of Gene fusion, placing a reporter gene to a regulatory sequence of a promoter. Blue bacteria of E. coli on a petri dish held by Jon Beckwith

Jon Beckwith described Malcolm in his letter to the family on September 21, 2009 (see Figure 2C):

> "Malcolm . . . was the most unusual person to ever work in my lab. He had his own Malcolm way of thinking. His mind was somewhere else. His mind was on science, but from a different perspective than many of us shared . . . It made me marvel all the more at his ingenuity. He conceived of clever steps, sometimes combining them all in the same reaction—something we would never do. And these strains and phages all worked and are still used widely in biology."

"He was in another world in some ways,
But what a world he created"

Beckwith was a scientist of great integrity, a true scholar, and an inspiring mentor. Malcolm often attributed his success to Beckwith for the values he established in his laboratory. In Beckwith's lab, Malcolm was cherished for his freestyle of thinking and nurtured by a kindred spirit. He was praised for his arduous love for science. Without the support and encouragement from Beckwith in his early career, Malcolm would not have so readily achieved the early level of excellence that brought him worldwide recognition.

Construction of Gene Fusion

In 1976, Malcolm came to Stanley Cohen's laboratory in Stanford with the goal of refining his gene fusion technology to become a broader-based generalized system. He intended to replace the complicated phage genetics with the easy manipulation of cloning *in vitro* on a plasmid that had just set pace in Stanley Cohen's laboratory. Malcolm had the vision to extend his fusion concept to a higher order of eukaryotic systems in yeast, plants, and animal cell lines, and ultimately in live organisms and animals throughout the biological kingdom. He had hoped to develop a generalized gene fusion system to study gene expression, gene regulation, and developmental specificity in live animals and humans. Of course, he didn't expect to do it all alone, but with many molecular biologists who came after him.

Malcolm's vision and scientific genius amazed many of us at Stanford. As a fresh Ph.D. from Harvard, Malcolm had laid out his vision of how to apply molecular biology to his science for the next thirty years. In only a few years, the concept and his gene fusion technology had become undergraduate and graduate class material encompassed in many standard biology textbooks and journals. Scientists and researchers rushed to apply the technology to their favorite genes under study. At the same time, Malcolm received highest honors and accolades from his contemporaries wherever he went. His invention, its application and usefulness, had expanded into many other biological systems far beyond what Malcolm and Beckwith had anticipated in 1975.

Along with the original gene fusion system, variations to chimeric protein fusion products had also been developed in early days. Chimeric proteins that maintain separate functionality of the two juxtaposed protein in a single protein allows the protein to be identified, purified, and earmarked in the cell while performing its original functions. Because Malcolm's original ideas of gene fusion and protein fusion systems were unpatented in 1975, few now remember his early endeavors in the development of gene fusion technology and their significant contribution to science.

An article in the *Journal of Bacteriology,* written by Jon Beckwith, Tom Silhavy, and Olaf Schneewind after Malcolm's death, with an introduction by the journal's editor, Dr. Philip Matsumura, Editor in Chief, chronicled the creativity of Malcolm's technology and gave an inside look at how gene fusion revolutionized the way scientists do gene experiments. (See Figure 2D, http://www.ncbi.nlm.nih.gov/pmc/articles/PMC2937382/).

> "Malcolm J. Casadaban died on 13 September 2009 from an infection and was found to have a weakened strain of the bacterium of Yersinia pestis in his blood. The tragic event took the life of one of the most creative and influential geneticists of our time. In the late 1970s and '80s, Malcolm invented novel approaches, which changed the way many of us did science. Jon Beckwith, Tom Silhavy, and Olaf Schneewind have chronicled his scientific life from graduate school to his death and give us insight into Malcolm's genius." (Editorial, Figure 2D)

Gene Fusion and The Nobel Prize

What left us to marvel was the transformation of a technology originated with a blue or red bacterial colony expressing ß-galactosidase on a petri dish to the ingenious demonstration of green florescent jellyfish shown in Figure 2C. In this experiment, GFP gene that encodes green florescence protein was used as a marker, a reporter gene and a tag juxtaposed to a jellyfish promoter on one end and the structural component of another protein in a gene/protein fusion system. The constructed DNA fusion was then placed back into the organism to allow green florescence to express in live jellyfish. The newly created jellyfish that expresses GFP protein *in vivo* allows the entire organism to shine in florescence under light as shown in the Figure 2C. This work culminated in a Nobel Prize in Chemistry in 2008, awarded by Royal Swedish Academy of Sciences to Osamu Shimomura of Boston University Medical School, Martin Chalfie of Columbia University, and Roger Y. Tsien of the University of California, San Diego, for making florescent jellyfish glow green. Gene fusion was indeed a tool to transform molecular biology into living cells and to confer specificity and design to the organisms by the creator who studied its gene expression and regulation. Based on the concept of gene fusion in 1975, Dr. Chalfie showed how green florescence could be used as a biological identifier, a reporter gene or tag by inserting green florescence into the body of a live jellyfish that allowed florescence expression under light (Figure 2E).

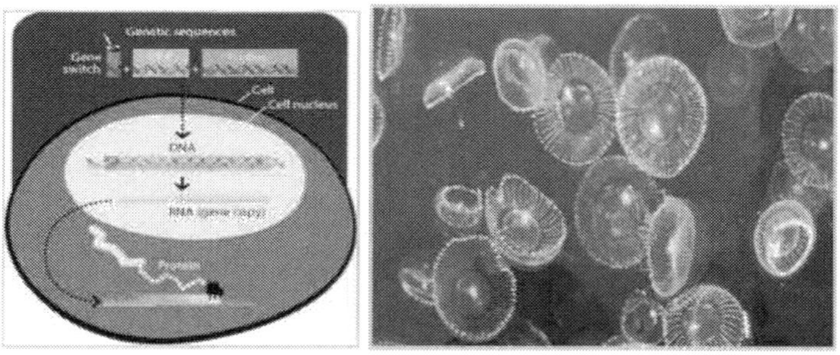

Figure 2E: Construct encoding GFP tagged protein harbored inside of Jellyfish allows the organism to glow in green florescence under light

Some biologists used the tool of green fluorescent reporter tag to track the growth and fate of specific cells, such as nerve cells damaged by Alzheimer's disease. The technique can even track specific proteins within cells. In one experiment, the brain of a mouse was transformed into a kaleidoscope of color by tagging different nerve cells with different fluorescent proteins.

Ironically, Malcolm did not share the glory of the Nobel Prize in 2008, but his gene fusion technology made it possible, as so many scientists, including the ones who earned the Nobel Prize, built upon it and gained knowledge about what happened inside the living organism or animals in a kaleidoscope.

The plane jerked and brought me back to the present moment. I sat up and glanced out of the window, seeing only darkness. Very few passengers were awake in the dim light of the plane's cabin. Most were quietly sleeping or reading a book. Even the flight attendants were quiet in the back.

Malcolm and I before Golden Gate Bridge at San Francisco

I wondered where Malcolm was at this hour. My stomach clutched with pain. Was he in that cold refrigerator in the morgue waiting for us to bring him home? I wanted to touch him, hold his hand, and ask him to tell me how it all happened. Sadly, I tipped my seat back again and reaching for him in the only way I could, in the memories of our life together.

A Fusion of Our Lives

Malcolm and I were married in 1977 in a grand mosaic Stanford Church on that landmark campus. We had a traditional Catholic wedding ceremony presided over by Malcolm's uncle, a priest of New Orleans. I wore a white lacy wedding gown with a long train and held a bouquet of red roses. Malcolm had cut his bushy hair, trimmed his beard, and wore a good-looking tuxedo. Both sets of our parents dressed in their finest. Our reception that evening was held at the Old Chef Chu's restaurant down the road from Stanford. Szechuan cuisine was the favorite among many Stanford professors and students in 1977. It seemed like yesterday that Malcolm and I took the first step in our lives together.

We were married on June 19, 1977 in Stanford church. Left were my parents and right were Malcolm's parents.

Our partnership extended into our professional lives. Malcolm and I had our first major joint paper published in the journal *Nature* in 1979 (Figure 2F). Tn3 is a genetic element that renders host bacterial cells resistant to the antibiotics of penicillin and ampicillin. The paper noted for the first time that Tn3 genetic element specified a novel enzyme, transposase, which mediates Tn3 transposition between DNA segments in an illegitimate recombination process. We had identified the transposition product within the Tn3 elements as well as the genetic circuits of a repressor regulated DNA transposition event between two DNA segments.

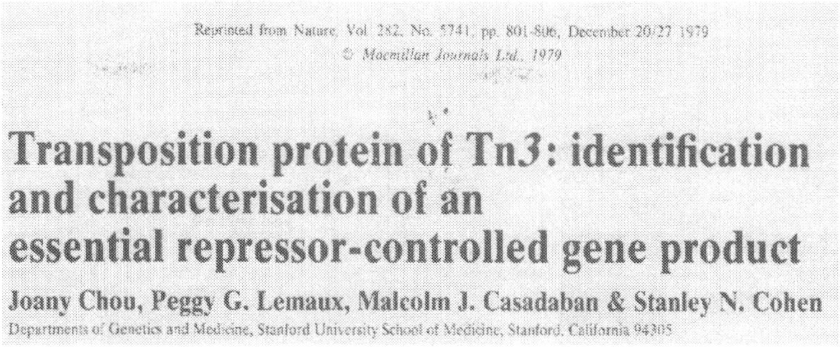
Reprinted from Nature, Vol. 282, No. 5741, pp. 801-806, December 20/27 1979
© Macmillan Journals Ltd., 1979

Transposition protein of Tn*3*: identification and characterisation of an essential repressor-controlled gene product

Joany Chou, Peggy G. Lemaux, Malcolm J. Casadaban & Stanley N. Cohen

Departments of Genetics and Medicine, Stanford University School of Medicine, Stanford, California 94305

Figure 2F: Our first article on Tn3 transposition published in Nature 1979.

When Malcolm approached completion of his postdoctoral studies at Stanford, it was natural for him to begin his search for a faculty position. He was invited to the California Institute of Technology (Caltech) to give a seminar; there he met two senior professors, Jerome Vinograd (1913–1976) and Melvin Simon, who took an interest in him and promised him a faculty position once he finished his postdoctoral study at Stanford. Malcolm was elated by the offer and the prospect of a faculty position at Caltech was the dream job he had hoped for since he left Harvard. He loved the science and reputational environment at Caltech, and he loved Pasadena, set in the heart of Los Angeles.

But fate seemed to have twisted him down a different path. Before Malcolm was able to finish his work at Stanford, Professor Jerome Vinograd passed away unexpectedly, leaving Malcolm to search for a faculty position elsewhere. Bob Haselkorn, who headed the Department of Biophysics and Theoretical Biology at the University of Chicago, approached Malcolm and offered him a faculty position in 1979. Malcolm accepted. This was his first job and the last in his lifetime. We could never have imagined how much strife and turmoil was in store for us or that death would stalk Malcolm there and finally strike him down on September 13, 2009, when he breathed his last on a bed at the University of Chicago hospital.

Chicago in the Early Years

On January 1, 1980, Malcolm and I drove into the Windy City to begin the next thirty years of our lives. A young married couple with all our worldly possessions packed up in our little Honda, we were bursting with excitement. We didn't even mind the piercing ice-cold wind that cut right through us when we drove into the city in the midst of a winter storm.

Brooke at age 3

Our first year in Chicago went by quickly. We had our first child, Brooke, in January of 1981. Brooke had blond hair, far more blond than any of the Caucasians I knew. Obviously, her father's gene was at work. Jokes surrounded us that Brooke was just another gene fusion product that Malcolm had created. Moments of our lives were bound together so long ago, and the vivid memory brought me back to the time when our daughter was born on the wintry night of January 7, 1981.

I had gone into labor earlier that morning at local Billings Hospital across from Midway. In the midst of intense pain right before the delivery, I slipped into a state of preeclampsia with my heart pounding so out of control and the pain so severe I couldn't breathe, I thought I was going to die at that moment. Several white coats were watching over me, and Malcolm was there. Although I couldn't see him, blinded as I was by the bright lights over me, I heard him cry out in his fear for my safety and that of our child. Dimly I heard the doctors rushing him out of the room to avoid unnecessary agitation. I must have

Malcolm with daughter Brooke

passed out after that. When I woke up, I saw Malcolm sitting next to me holding our first child in his arms, beaming with relief and the pride of a new father.

Tom McGarry remembers Malcolm

Tom McGarry, Professor at Northwestern University, had this to say about Malcolm:

> *"I knew him in the early 80's when I was a graduate student at the U of C and he was a new faculty member. I took a course that he taught in bacterial genetics and I rotated through his lab for a few months. This was around the time you were born. In fact, I remember that Dr. Haselkorn was a substitute teacher for the first class meeting because Malcolm was at the hospital for your birth. He came into the room that day with a serious look on his face and said, "You all know where Malcolm is?" and we nodded."*

And Tom continued.

> *"Your father was extremely intelligent; he was one of the smartest guys I ever met. He was really engaged - his phone rang constantly, and whenever we talked to him we would expect to be interrupted by several phone calls. I learned a lot from him. And his course had a major influence on my thinking and how I approach science. I remembered that at the start of his first lecture he drew an oval with a circle inside it on the blackboard and said, "This is our model of the cell – a bag with a piece of DNA in it." I thought, "I like the way this guy thinks." He was also very friendly and approachable, always in a good mood, laughing and joking all the time. .... I really like him a lot. I remember that when I went to ask him if I could work in his lab, he said he remembered me from class and thought I was a smart guy who asked good questions. Coming from him, that was quite a compliment. ...."*

> *"He made a huge contribution to the field of bacterial genetics. The whole idea of gene fusions to study gene expression started with him. The technology is used worldwide nowadays and*

continues to evolve – enhancer traps and luciferase reporters are all derived from what he did. You should be proud of him. But most of all he was just a wonderful guy. I share your sadness now that he's gone."

In 1987, our second child was born. Leigh was very smart even as a small child and wanted to be a surgeon someday. She would follow her father's footstep in science to his alma mater at MIT in Boston.

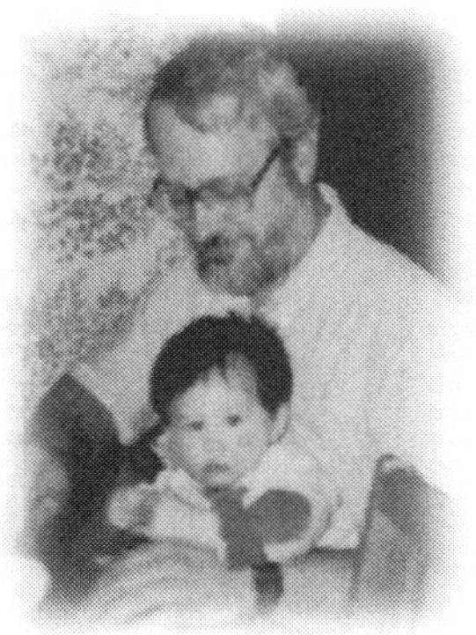

Malcolm and Leigh

Although a renowned scientist and acknowledged intelligence, Malcolm led a simple life, devoted to his science and his family. He was a man of unusual kindness and gentleness of spirit, who spoke softly—so softly that people often had to lean forward to hear him—and, for all his experience, seemed rather unworldly.

As Jon Beckwith wrote,

"Malcolm was a genius and he was a naïve, at times a childlike person in an endearing (if sometimes frustrating) way." (Figure 2C)

This was the Malcolm I knew and who he really was.

Through the years, Malcolm had several students in the laboratory, including Stuart Shapira, Alfonso Martinez-Arias, Mary Ditto, and Jonathan Kans and many others whom I got to know. Many of them had since attained high posts in academia, government, and the private sector. One student, Sagar Koduri, wrote of Malcolm:

Researcher remembered

I grieved to hear of Malcolm Casadaban's passing last month (see Deaths*). For those of us in the mid-1990s who were taking our first undergraduate steps into the world of molecular biology, Malcolm was an indelible figure. He certainly changed my life and propelled me into choosing life as a professional scientist.*

As a freshman looking for a summer research project and all of 17, I knew next to nothing about molecular genetics—and at our first meeting was utterly unprepared for this mild-mannered man with bright blue eyes and an uncombed shock of crazy white hair. And even less prepared for a rambling, chaotic, joyous, tumultuous, full-on conversation on the genetic architecture of the arabinose, rhabdose, rhamnose *operons, the possibility of adapting them for use as tools for the detection of protein-protein interactions, preliminary experiments using bacteriophages to create the relevant constructs, and somehow touching on transgenic pigs—that lasted for three hours and was utterly over my head and felt like being trapped inside a fire hydrant. And I loved it.*

Over the years that I worked with Malcolm, I learned a tremendous amount about bacteria, phages, and molecular biology. And I did so without the benefit of an organized textbook, by talking to someone who quite literally helped invent the field (though it was only from others that [I] learned just how famous he was—that was the essential modesty of the man). I will always remember him as the most brilliant man I've met—bursting at the seams with ideas, thoughts, and dreams. He was a nightmare to TA for—it involved being a cross between Xanthippe and a sheepdog to try and corral him through a simple genetics demonstration because even there, he just wanted to talk, and think, and dream about science, to share his fascination and his delight.

And it was never even necessarily his own work—his office (a surreal landscape of towering stacks of paper that were five- and six-feet tall) was littered with emails and notes from colleagues thanking him for the time, effort, and generosity with which he listened to others' scientific puzzles and offered ingenious, creative, and brilliant advice. Because more than anything, Malcolm loved science. Loved it the way that some kids love baseball or football or cricket. But loved it and rejoiced in it because it was play to him. Not a game, I want to say— there wasn't anything trivial about it. Play, as they say, for mortal stakes.

I am, many years later, about to start my own postdoctoral fellowship. I don't think I'll ever be lucky enough to have a brain like Malcolm's. Malcolm dreamed in riotous Technicolor: so active, popping with wonderful new ideas, and able to extrapolate with his unique blend of rigor and verve, from an interesting fact or two to an incisive and beautifully elegant research problem. I am a far more pedestrian scientist. But I hope that all of us who were lucky enough to know him will carry on something of his kindness, his utter humility, his extraordinary gentleness, and above all else, that playful, brilliant, leaping delight in science. We are all poorer for his passing, and I know I will miss him.

Sagar Koduri, SB'97
Brookline, Massachusetts
[Koduri, S. Letters, *University of Chicago Magazine*, Nov–Dec 2009]

This is who Malcolm really is, as Koduri artfully put it, *"Because more than anything, Malcolm loved science. Loved it the way that some kids love baseball or football or cricket. But loved it and rejoiced in it because it was play to him. Not a game, I want to say— there wasn't anything trivial about it. Play, as they say, for mortal stakes".*

Malcolm and his students in 1985 at our house: Stuart Shapira, Alfonso Martinez-Arias, Mary Ditto, Dave Nielson, me and Malcolm

Mortal stakes, indeed. Malcolm—brilliant, generous, kind, fun loving, and naïve—had no idea that the stakes would turn out to be so high for him at Chicago. None of us did.

Malcolm and I attended a meeting in Stanford, 2005

HARVARD MEDICAL SCHOOL
Department of Microbiology
and Molecular Genetics

200 LONGWOOD AVENUE
BOSTON, MASSACHUSETTS 02115

(617) 432- 1920
FAX: (617) 738-7664

September 21, 2009

Department of Microbiology at Cummings Life Sciences Center
920 East 58th Street
Chicago, IL 60637

To the family of Malcolm Casadaban,

I was terribly saddened to hear of Malcolm's sudden death. It brought up in me the deep fondness I felt for him and profound appreciation of all he done for me and my colleagues.

Within a few hours of hearing of Malcolm's death, I realized something I had put together in my mind of what Malcolm had accomplished while he worked with me and in Stanley Cohen's lab. Much of what we do in my lab today, AND what other people do in biology, we owe to Malcolm. In a series of papers from my lab, Malcolm presented to the biological world generalized methods for in vivo construction of gene fusions. Even though the recombinant DNA era brought in a new technology for doing it, Malcolm's approach was copied and altered for new purposes, and is still in use today. Importantly, it introduced the idea of scanning the chromosome by gene fusions which has been applied to many problems, including widely to bacterial pathogenesis.

Malcolm was a genius and he was a naïf, at times a childlike person in an endearing (if sometimes frustrating) way. He was the most unusual person to ever work in my lab. He had his own Malcolm way of thinking. His mind was somewhere else. His mind was on science, but from a different perspective then many of us shared. He loved to construct. If I remember correctly, he once told me that he grew up watching his father who was a carpenter and that explained why he liked to construct. He constructed a set of bacterial strains and phages that are still widely used in biology- MC1000 and MC4100, for two. We are often reminded of him (and his personality) simply when we go to dig out one of these (and other) strains of his. People have asked me a number of times about the origins of some of his strains. I went back to our records and tried to trace back these strains, and after going through from 15 to 20 precursor strains that Malcolm had constructed, I gave up. I thought I would have to give up a weekend to do it. It made me marvel all the more at his ingenuity. He conceived of clever steps, sometimes combining them all in the same reaction- something we would never do. And these strains and phages all worked and are still used widely in biology. When he did this work, it was so clearly all his own, that I took the highly unusual step in my field of letting him publish three of his classic papers under his own name.

One of his great qualities, other than being a genius, was that he loved to help people. In my lab, at conferences, on the phone- he tutored people in the intricacies of his strains that were so useful. It might take some time, because when he would begin talking, he would start in the middle of thoughts instead of the beginning. What an interesting brain he must have had.

He was in another world in some ways, but what a world he created.

One last story: Malcolm was very soft-spoken- sometimes hard to hear; maybe also there was some remnants of a Louisianan twang. He appeared, in that sense, to be a very meek person. But, coming to attend a recent reunion of people who had worked in my lab, in Amalfi, Italy, Malcolm had his wallet stolen while walking around Rome. He immediately set chase after the pickpocket and recovered his wallet. He was certainly not the threatening type, but he must have shown something, that I was unaware of in that adventure.

You have had a pretty special and unusual person in your family.

Sincerely yours,

Jonathan Beckwith
American Cancer Society Professor

Figure 2C: Letter from Jonathan Beckwith to the Family

Journal of Bacteriology, Sept. 2010, p. 4261–4263 Vol. 192, No. 17
0021-9193/10/$12.00 doi:10.1128/JB.00484-10

EDITORIAL

Remembering Malcolm J. Casadaban▽

Malcolm J. Casadaban died on 13 September 2009 from an infection and was found to have a weakened strain of the bacterium *Yersinia pestis* in his blood. This tragic event took the life of one of the most creative and influential geneticists of our time. In the late 1970s and '80s, Malcolm invented novel approaches which changed the way many of us did science. Jon Beckwith, Tom Silhavy, and Olaf Schneewind have chronicled his scientific life from graduate school to his death and give us insight into Malcolm's genius.

Philip Matsumura
Editor in Chief, Journal of Bacteriology

IN MEMORIAM

Malcolm J. Casadaban 1949–2009

Malcolm J. Casadaban, who passed away on 13 September 2009, was an imaginative experimentalist whose technological and intellectual innovations were used by a generation of scientists employing genetic approaches in microbes. Malcolm's methods even transcended the typical divide that separates investigators of microbial and higher eukaryotic systems.

Born in New Orleans in 1949, Malcolm was the first of seven children raised by John and Dolores Casadaban. After earning his diploma from Jesuit High School in his home town, he studied biology at the Massachusetts Institute of Technology (MIT) and graduated in 1971. Inspired by his teachers at MIT, Malcolm pursued graduate training at Harvard Medical School, where he joined the laboratory of one of us (J.B.) and completed his Ph.D. requirements in 1976. Even though he was only at the beginning of his career, this gentle, naïve young man revolutionized genetic approaches to studying a host of biological problems. Starting in the early 1970s, in a series of papers from the Beckwith laboratory (4-6) and subsequently with his postdoctoral fellowship mentor (10, 11), Stanley Cohen, at Stanford University, Malcolm presented to the biological world increasingly sophisticated methods for construction of gene fusions *in vivo*. Even though the recombinant-DNA era later brought in a new *in vitro* technology that allowed construction of fusions to genes individually, Malcolm's approach still had broad uses. Importantly, it allowed scanning of chromosomes by using gene fusions, a technique that has been applied to many problems (37).

Malcolm's graduate mentor (J.B.) recalls his unusual personality. He loved to construct bacterial strains with new properties, bringing together complex, almost architectural plans. Malcolm once explained that his interest in genetic constructions originated from his youth, as he grew up watching his father work as a carpenter. Malcolm's strain constructions sometimes involved steps that went against tradition, for instance, mixing together transduction and conjugation steps in one tube. Tracing the origins of widely used Casadaban strains such as MC1000 and MC4100 can be a rapidly confusing enterprise because he involved 20 or more precursor strains (6). These strains also illustrate Malcolm's ingenuity and the nature of his work products, which were so clearly his own. For these reasons, Jon Beckwith told Malcolm that he should publish the papers reporting his graduate work on gene fusion techniques under his name alone (4-6).

Gene fusion is used so commonly now that the origins of this powerful technology have become obscure. Before Malcolm, the number of existing gene fusions could be counted on the fingers of one hand. They were constructed by giants in the field of bacterial genetics using cumbersome multistep procedures that were difficult to generalize (2, 24). Malcolm devised a three-step procedure for constructing gene fusions that could be generalized for any nonessential gene in *Escherichia coli*. At the heart of Malcolm's method is a plaque-forming λ *lac*-specialized transducing phage called λp1(209) (6). It is important to note that this phage was constructed with toothpicks (36); no recombinant-DNA methodology was involved. To appreciate Malcolm's genius, the reader is encouraged to look at the paper describing this phage construction (6).

The first step in Malcolm's procedure was to isolate a Mu insertion in the gene of interest in a strain like MC4100, which carries a deletion of the *lac* operon (6). This strain is then lysogenized with λp1(209). Since λp1(209) lacks an attachment site, it can integrate into the chromosome only by homologous recombination between the Mu DNA elements, and this recombination places the *lac* operon within the gene of interest. Provided that the Mu prophage is integrated in the proper orientation, it also places the *lac* operon downstream from the promoter of the gene of interest. Since the Mu prophage carries a temperature-sensitive mutation in the repressor gene, selection for the Lac^+ phenotype at high temperatures will yield survivors that have suffered a deletion event that removes the lethal Mu prophage and fuses the *lac* operon to the promoter of the gene of interest (6). Fusions isolated by this method are called operon or transcriptional fusions, and researchers in many labs took advantage of Malcolm's method to study their particular gene or regulatory system of interest

▽ Published ahead of print on 28 May 2010.

(37). For example, *malP-lacZ*$^+$ fusions were used to isolate the first constitutive mutations in the regulatory gene *malT*, and these mutations were used to show that MalT regulates the *mal* operons in a strictly positive fashion (15). Without fusions, there was no way to isolate such constitutive mutations because chemists had not been able to synthesize noninducing substrates of the *mal* regulon. With fusions, lactose becomes a noninducing substrate for any regulatory system. Simply selecting for the Lac$^+$ phenotype with these fusion strains yields constitutive mutations.

Malcolm cleverly modified his three-step method so that it could be used to isolate true gene fusions, i.e., fusions that create a hybrid gene that specifies a hybrid protein with functional LacZ at the carboxy terminus. He did this by introducing a nonsense mutation at codon 18 of *lacZ*, producing the specialized transducing phage λp1(209,118) (6). Of course, this was also done with toothpicks. Now to get Lac$^+$ survivors at high temperatures, the deletion that removes the lethal Mu prophage must also enter the *lacZ* gene and remove the nonsense mutation. The target for the end point of this deletion is very small; the deletion must remove the nonsense mutation, but it can't extend much further into *lacZ* without destroying LacZ function. Accordingly, survivors are rare, but these fusion strains proved to be of tremendous value. For example, they opened the door for genetic analysis of protein secretion. When the amino terminus of a periplasmic protein such as MalE or an outer membrane protein such as LamB is fused to LacZ, the cell attempts to secrete the hybrid protein (1) (38). It turns out that LacZ doesn't fold properly if secreted from the cytoplasm, and the unexpected Lac$^-$ phenotype of such fusion strains allowed the isolation of the first signal sequence mutations (18) and the identification of genes that specify components of the cellular protein secretion machinery (17, 31).

There are many pitfalls in Malcolm's three-step method, and consequently, many who wanted to use gene fusion technology were reluctant to try it. So as a postdoctoral fellow in Stanley Cohen's laboratory, Malcolm developed a one-step method (11). For this method, Malcolm inserted a promoterless *lac* operon, together with the gene for β-lactamase (penicillin resistance), within the Mu genome to produce a defective specialized transducing phage, Mu*d*(Ap, *lac*). Malcolm used recombinant DNA in the construction of this phage, but only at one of many steps. One of the intermediates in this construction would be very rare and extremely difficult to identify. In typical Malcolm fashion, he never did identify it. He simply assumed it must be there and carried on. Clearly, this intermediate must have been there! His report is another paper that provides real insight into his genius (11). In a subsequent paper, Malcolm and his wife, Joany Chou, describe a related Mu*d*(Ap, *lac*) phage that can be used to construct fusions that make LacZ hybrid proteins (7).

The Mu*d*(Ap, *lac*) phages were the first of the fusion-generating transposons. With these elements, constructing fusions is really simple (7, 11). You infect a desired strain with the Mu*d*(Ap, *lac*) phage, for example, wait 20 min, and then plate the infected cells onto medium that selects for the loss of a gene of interest and contains ampicillin and a LacZ color indicator such as X-Gal (5-bromo-4-chloro-3-indolyl-β-D-galactopyranoside). Then you go have a beer, watch a football game, and come back in the morning to purify your fusions (the blue colonies). It is true that modern technologies such as PCR have greatly simplified the construction of gene fusions to known genes. However, fusion-generating transposons have additional uses. For example, they can be used to find unknown genes that belong to a particular regulon. This is done simply by looking for fusions that are regulated in a particular way. In this manner, the genes that specify the proteins that comprise the redundant arabinose transport systems were identified (26). This could not be done in traditional ways because mutants lacking only one of the transport systems have no relevant phenotype. Similarly, the transposons were used to identify unknown genes regulated by the SOS response (25). Nowadays, fusions are commonly employed and there are many reporters other than LacZ, with green fluorescent protein (GFP), the discovery of which led to a Nobel Prize, being just one example (13). But we should not forget that it was Malcolm's work that popularized this experimental approach and first made it generally available.

In 1980, Malcolm accepted a faculty appointment at the University of Chicago. One of many immediate research foci was the characterization of *E. coli* Tn*3* transposase and its regulation (9). Once again, Malcolm was aiming to develop facile tools for transposon mutagenesis (8, 14, 16). Other work expanded the uses of gene fusion in *E. coli* and examined the universality of this technology for other systems, a purpose that led Malcolm to express β-galactosidase in yeast (12, 35). Malcolm's first graduate student, Alfonso Martinez-Arias, used *lacZ* fusions to isolate the upstream regulatory element of the yeast *leu2* gene, which enabled molecular studies on transcriptional repression by leucine at this promoter (27, 28). In a collaboration between Malcolm's and Donald Steiner's laboratories, Malcolm's *lacZ* fusion technologies enabled the first study of insulin gene expression in pancreatic β-cells (29, 30).

Genetic technology was developed further by his graduate student Eduardo Groisman, who created mini-Mu phages carrying a plasmid replicon and antibiotic resistance cassettes for *in vivo* cloning (22, 23). The power of this technology was revealed through the rapid isolation of mutants with mini-Mu transposon insertions and measurements of gene expression via *lacZYA* fusion. Desired mutations could then be studied by transducing loci into other genetic backgrounds by using Mu helper phages or rapid recombination cloning of gene fusions via the associated plasmid replicons. This technology was expanded for λ helper phages and cosmids, providing a suite of tools for rapid generation, isolation, and transfer of mutations in *E. coli* and other Gram-negative bacteria (20, 21).

By 1990, Malcolm directed his research to address the need for genetic tools for Gram-negative pathogens, initially studying *Pseudomonas aeruginosa* and its transposable bacteriophages and then *Salmonella enterica* serovar Typhimurium, in which he identified core lipopolysaccharide genes to demonstrate the use of his tools (32–34). At the same time, Malcolm was engaged in founding and developing a biotechnology company, ThermoGen Inc., exploring the genes of the archaeon *Thermus flavus* for thermostable enzymes with industrial uses (40, 41). This work was carried forward by David C. Demirjian, Malcolm's last Ph.D. student.

In pursuit of new research programs, Malcolm collaborated with one of us (O.S.) on the type III secretion pathway of pathogenic *Yersinia* species. Malcolm initially developed tools for transposon mutagenesis as well as gene transfer and mapping. He eventually set out to identify mutations that affect the low-calcium responses of these microbes. Briefly, chelation of calcium ions activates type III secretion, and this arrests the growth of yersiniae (19). Translational hybrids of secretion substrate genes were generated as *lacZ* fusions, and the resulting proteins can block type III secretion, enabling the selection of mutations that abrogate substrate recognition and suppression of those mutations (3, 39). In the midst of these experiments, Malcolm unexpectedly passed away, his last studies still unfinished.

Already as a graduate student, Malcolm was an endearingly, and sometimes frustratingly, naïve person. He might absorb and mention something about some news item or some political perspective, but it was not something that he was going to put a lot of thought into. His mind was fully involved with his science, his constructions. Malcolm was very soft-spoken, with maybe the remnants of a Louisianan twang that made him occasionally hard to hear. Yet he loved to help people. In the laboratory, at conferences, or on the phone—he tutored the many people who were using his strains about their intricacies. He clearly loved to teach and was unstintingly ready to instruct, however long it took.

Malcolm J. Casadaban's ingenuity at inventing genetic tools and his dedication to experimental work on complex biological problems will continue to serve as an inspiration for the scientific community. He is survived by his parents, brothers, and sisters; by his two daughters, Brooke and Leigh, and his former wife, Joany Chou; and by his fiancée, Casia Holmgren.

We acknowledge the help of Bill Blaylock, Eduardo Groisman, and Robert Haselkorn in preparing the manuscript.

REFERENCES

1. **Bassford, P. J., Jr., T. J. Silhavy, and J. Beckwith.** 1979. Use of gene fusion to study secretion of maltose-binding protein into *Escherichia coli* periplasm. J. Bacteriol. **139:**19–31.
2. **Beckwith, J. R., E. R. Signer, and W. Epstein.** 1966. Transposition of the *Lac* region of *E. coli*. Cold Spring Harbor Symp. Quant. Biol. **31:**393–401.
3. **Blaylock, B., J. A. Sorg, and O. Schneewind.** 2008. *Yersinia enterocolitica* type III secretion of YopR requires a structure in its mRNA. Mol. Microbiol. **70:**1210–1222.
4. **Casadaban, M. J.** 1975. Fusion of the *Escherichia coli lac* genes to the *ara* promoter: a general technique using bacteriophage Mu-1 insertions. Proc. Natl. Acad. Sci. U. S. A. **72:**809–813.
5. **Casadaban, M. J.** 1976. Regulation of the regulatory gene for the arabinose pathway, *araC*. J. Mol. Biol. **104:**557–566.
6. **Casadaban, M. J.** 1976. Transposition and fusion of the *lac* genes to selected promoters in *Escherichia coli* using bacteriophage lambda and Mu. J. Mol. Biol. **104:**541–555.
7. **Casadaban, M. J., and J. Chou.** 1984. *In vivo* formation of gene fusions encoding hybrid beta-galactosidase proteins in one step with a transposable Mu-lac transducing phage. Proc. Natl. Acad. Sci. U. S. A. **81:**535–539.
8. **Casadaban, M. J., J. Chou, and S. N. Cohen.** 1982. Overproduction of the Tn3 transposition protein and its role in DNA transposition. Cell **28:**345–354.
9. **Casadaban, M. J., J. Chou, P. G. Lemaux, C. P. Tu, and S. N. Cohen.** 1981. Tn3: transposition and control. Cold Spring Harbor Symp. Quant. Biol. **45:**269–273.
10. **Casadaban, M. J., and S. N. Cohen.** 1980. Analysis of gene control signals by DNA fusion and cloning in *Escherichia coli*. J. Mol. Biol. **138:**179–207.
11. **Casadaban, M. J., and S. N. Cohen.** 1979. Lactose genes fused to exogenous promoters in one step using a Mu-lac bacteriophage: *in vivo* probe for transcriptional control sequences. Proc. Natl. Acad. Sci. U. S. A. **76:**4530–4533.
12. **Casadaban, M. J., A. E. Martinez-Arias, S. K. Shapira, and J. Chou.** 1983. Beta-galactosidase gene fusions for analyzing gene expression in *Escherichia coli* and yeast. Methods Enzymol. **100:**293–308.
13. **Chalfie, M., Y. Tu, G. Euskirchen, W. Ward, and D. Prasher.** 1994. Green fluorescent protein as a marker for gene expression. Science **263:**802–805.
14. **Chou, J., P. G. Lemaux, M. J. Casadaban, and S. N. Cohen.** 1979. Transposition protein of Tn3: identification and characterization of an essential repressor-controlled gene product. Nature **282:**801–806.
15. **Debarbouille, M., H. A. Shuman, T. J. Silhavy, and M. Schwartz.** 1978. Dominant constitutive mutations in *malT*, the positive regulator of the maltose regulon in *Escherichia coli*. J. Mol. Biol. **124:**359–371.
16. **Ditto, M. D., J. Chou, M. W. Hunkapiller, M. A. Fennewald, S. P. Gerrard, L. E. Hood, S. N. Cohen, and M. J. Casadaban.** 1982. Amino-terminal sequence of the Tn3 transposase protein. J. Bacteriol. **149:**407–410.
17. **Emr, S. D., S. Hanley-Way, and T. J. Silhavy.** 1981. Suppressor mutations that restore export of a protein with a defective signal sequence. Cell **23:**79–88.
18. **Emr, S. D., M. Schwartz, and T. J. Silhavy.** 1978. Mutations altering the cellular localization of the phage lambda receptor, an *Escherichia coli* outer membrane protein. Proc. Natl. Acad. Sci. U. S. A. **75:**5802–5806.
19. **Goguen, J. D., J. Yother, and S. C. Straley.** 1984. Genetic analysis of the low calcium response in *Yersinia pestis* Mu d1(Ap *lac*) insertion mutants. J. Bacteriol. **160:**842–848.
20. **Groisman, E. A., and M. J. Casadaban.** 1987. Cloning of genes from members of the family *Enterobacteriaceae* with mini-Mu bacteriophage containing plasmid replicons. J. Bacteriol. **169:**687–693.
21. **Groisman, E. A., and M. J. Casadaban.** 1987. In vivo DNA cloning with a mini-Mu replicon cosmid and a helper lambda phage. Gene **51:**77–84.
22. **Groisman, E. A., and M. J. Casadaban.** 1986. Mini-Mu bacteriophage with plasmid replicons for in vivo cloning and *lac* gene fusing. J. Bacteriol. **168:** 357–364.
23. **Groisman, E. A., B. A. Castilho, and M. J. Casadaban.** 1984. *In vivo* DNA cloning and adjacent gene fusing with a mini-Mu-lac bacteriophage containing a plasmid replicon. Proc. Natl. Acad. Sci. U. S. A. **81:**1480–1483.
24. **Jacob, F., A. Ullman, and J. Monod.** 1965. Deletions fusionnant l'operon lactose et un operon purine chez *Escherichia coli*. J. Mol. Biol. **13:**704–719.
25. **Kenyon, C. J., and G. C. Walker.** 1980. DNA-damaging agents stimulate gene expression at specific loci in *Escherichia coli*. Proc. Natl. Acad. Sci. U. S. A. **77:**2819–2823.
26. **Kolodrubetz, D., and R. Schleif.** 1981. L-Arabinose transport systems in *Escherichia coli* K-12. J. Bacteriol. **148:**472–479.
27. **Martinez-Arias, A. E., and M. J. Casadaban.** 1983. Fusion of the *Saccharomyces cerevisiae leu2* gene to an *Escherichia coli* beta-galactosidase gene. Mol. Cell. Biol. **3:**580–586.
28. **Martinez-Arias, A. E., H. J. Yost, and M. J. Casadaban.** 1984. Role of an upstream regulatory element in leucine repression of the *Saccharomyces cerevisiae leu2* gene. Nature **307:**740–742.
29. **Nielsen, D. A., J. Chou, A. J. MacKrell, M. J. Casadaban, and D. F. Steiner.** 1983. Expression of a preproinsulin-beta-galactosidase gene fusion in mammalian cells. Proc. Natl. Acad. Sci. U. S. A. **80:**5198–5202.
30. **Nielsen, D. A., M. Welsh, M. J. Casadaban, and D. F. Steiner.** 1985. Control of insulin gene expression in pancreatic beta-cells and in an insulin-producing cell line, RIN-5F cells. I. Effects of glucose and cyclic AMP on the transcription of insulin mRNA. J. Biol. Chem. 260-13585-13589.
31. **Oliver, D. B., and J. Beckwith.** 1981. *E. coli* mutant pleiotropically defective in the export of secreted proteins. Cell **25:**765–772.
32. **Roncero, C., and M. J. Casadaban.** 1992. Genetic analysis of the genes involved in synthesis of the lipopolysaccharide core in *Escherichia coli* K-12: three operons in the *rfa* locus. J. Bacteriol. **174:**3250–3260.
33. **Roncero, C., A. Darzins, and M. J. Casadaban.** 1990. *Pseudomonas aeruginosa* transposable bacteriophages D3112 and B3 require pili and surface growth for adsorption. J. Bacteriol. **172:**1899–1904.
34. **Roncero, C., K. E. Sanderson, and M. J. Casadaban.** 1991. Analysis of the host ranges of transposon bacteriophages Mu, MuhP1, and D108 by use of lipopolysaccharide mutants of *Salmonella typhimurium* LT2. J. Bacteriol. **173:**5230–5233.
35. **Rose, M., M. J. Casadaban, and D. Botstein.** 1981. Yeast genes fused to beta-galactosidase in *Escherichia coli* can be expressed normally in yeast. Proc. Natl. Acad. Sci. U. S. A. **78:**2460–2464.
36. **Shuman, H. A., and T. J. Silhavy.** 2003. The art and design of genetic screens: *Escherichia coli*. Nat. Rev. Genet. **4:**419–431.
37. **Silhavy, T. J., and J. R. Beckwith.** 1985. Uses of *lac* fusions for the study of biological problems. Microbiol. Rev. **49:**398–418.
38. **Silhavy, T. J., M. J. Casadaban, H. A. Shuman, and J. R. Beckwith.** 1976. Conversion of beta-galactosidase to a membrane-bound state by gene fusion. Proc. Natl. Acad. Sci. U. S. A. **73:**3423–3427.
39. **Sorg, J. A., N. C. Miller, M. M. Marketon, and O. Schneewind.** 2005. Rejection of impassable substrates by *Yersinia* type III secretion machines. J. Bacteriol. **187:**7090–7102.
40. **Vonstein, V., S. P. Johnson, H. Yu, M. J. Casadaban, N. C. Pagratis, J. M. Weber, and D. C. Demirjian.** 1995. Molecular cloning of the *pyrE* gene from the extreme thermophile *Thermus flavus*. J. Bacteriol. **177:**4540–4543.
41. **Weber, J. M., S. P. Johnson, V. Vonstein, M. J. Casadaban, and D. C. Demirjian.** 1995. A chromosome integration system for stable gene transfer into *Thermus flavus*. Biotechnology **13:**271–275.

Figure 2D: Phil Matsumura, Editor of Journal of Bacteriology, remembered Malcolm's Gene Fusion system and its creativity

3

CHANGE WINDS IN CHICAGO

In the dim light on the plane, my thoughts journeyed back to year 2000 on a patent dispute that resulted in a highly publicized civil suit against the University of Chicago and Bernard Roizman. It was then I left Chicago and the life Malcolm and I had built there together. The changing winds in Chicago had twisted up, darkened the sky, and transformed into the funnel of a tornado that would tear down our home in the following year.

In 1986 I earned a Ph.D. in virology from the University of Chicago and decided to stay as a postdoctoral fellow in the laboratory of Bernard Roizman, for whom I had worked as a graduate student. I was productive and competent in my research career over the next ten years and published a series of research articles on the discovery of a novel gene that I called $\gamma_1 34.5$. It was an important gene that imparted neurovirulence property to the herpes simplex virus in affliction to human that lasted for more than a thousand years.

I was the sole discoverer of $\gamma_1 34.5$ in Roizman's laboratory and for as many as years I studied it, I nearly memorized its entire coding sequence. So long and intensely did I dwell on its characterization and deciphering its function that the gene synced with my name and rhymed with my signature.

Chou v The University Of Chicago

In the mid-1990s, my scientific knowledge and research skills had reached the level at which I needed to strike out on my own, find a faculty position elsewhere, to establish myself in my own right as a scientist of the world. I went on several interviews but failed to get an academic position anywhere. People at some academic institutions who considered me for a position asked me how I was going to live apart from a spouse who had academic standing at the University of Chicago. Other people thought that I had been a fellow too long in Roizman's laboratory and that would hamper my independence in my research career.

Whatever the reason, although I spent nearly fifteen years of my life working for a mentor at an NIH fellowship salary, made important discoveries, and published numerous academic articles, I was not able to leave to move on to a suitable position to begin my own career. To make matters worse, people at the University of Chicago who were envious of my research success began to spread rumors about what I said and what I did and reported them to Roizman. My relationship to Roizman began to deteriorate, and one day in 1996 I was given the pink slip to leave. Bin He was given the project that I had been working on for the past fifteen years and he later took an academic position that I had interviewed with at the University of Illinois at Chicago.

As it turned out, while I was busily working in his laboratory, Roizman was pursuing a hidden agenda. What I discovered then compelled me to bring legal action against the University of Chicago and Bernard Roizman in an unprecedented court action.

A review of the case in the *Duke University Law & Technology Review*, titled "A Victory for the Student Researcher: Chou v. University of Chicago," summarized what happened in the laboratory and in the legal case:

> The plight of Dr. Joany Chou, a researcher in molecular genetics at the University of Chicago working under the supervision of Dr. Bernard Roizman, exemplifies the problems faced by many student researchers. Dr. Chou began working with Dr. Roizman in 1983 studying a potential vaccine for the herpes virus. After successfully completing the vaccine in February 1991, Dr. Chou expressed to Dr. Roizman her belief that the discovery should be patented and inquired as to the proper procedures to obtain patent protection. Dr. Roizman responded to this inquiry by stating that the discovery was not patentable.
>
> Trusting the wisdom and experience of her friend and mentor, Dr. Chou returned to her research and forgot all aspirations of obtaining a patent. Over the next five years, Dr. Chou and Dr. Roizman continued to enjoy research success, publishing various papers and jointly securing a patent on a distinct aspect of their herpes research. As the leader of this successful research group, Dr. Roizman received various awards and became known among his peers as one of the best researchers in the field.
>
> However, in June 1996, Dr. Roizman approached Dr. Chou and told her that he was going to fire her if she did not resign. . . . Confused at this sudden treatment by her mentor, Dr. Chou began investigating the situation and discovered that in 1991, . . . Dr. Roizman secretly filed a patent application on the vaccine that he had portrayed as unpatentable to Dr. Chou. This patent issued a few years later and declared Dr. Roizman as the sole inventor. Although the patent was now owned by the University of Chicago . . . , Dr. Roizman had been receiving substantial royalty payments from the University's licensing of the patented technology. . . . Dr. Roizman had forced Dr. Chou to resign because if she ever found out about the secret patent or the royalty payments, it would be much easier for her to prove that she indeed was the rightful inventor of the patent by

demonstrating her access to the laboratory files. [Kyle Grimshaw, "A Victory for the Student Researcher: Chou v. University of Chicago," 1 *Duke Law & Technology Review* 1–7 (2001)]

Feeling betrayed by the professor for whom I had worked fifteen years and had compiled more than thirty publications and several patents, his duplicity and outright lie, shocked and angered me. He had taken my invention for his personal benefit and then put me out on the street. I had been in academia for many years, and while I knew the system was far from perfect, I had still believed in the overall integrity of the university and its professors.

Appealing to University of Chicago officials brought no relief. Having no other recourse, I then brought civil action against the University of Chicago and Roizman. The case went from the U.S. District Court for the Northern District of Illinois (the trial court) to the U.S. Court of Appeals for the Federal Circuit. It was in the middle of April 2001, when cherry blossoms filled the air of Washington, D.C., that I went to listen to the determination of my case before the three appellate judges of the Federal Circuit.

The Lawsuit Settlement

The summary of my claims was as follows:

A. Standing to Sue for Correction of Inventorship under 35 U.S.C.§256

B. Declaratory Judgment to Correct Inventorship

C. State Law Claims

1. Fraudulent Concealment
2. Breach of Fiduciary Duty
3. Unjust Enrichment
4. Breach of Express Contract
5. Breach of Implied Contract
    a. Implied-in-fact Contract
    b. Implied-in-law Contract

D. Miscellaneous

1. Academic Theft and Fraud
2. Case Reassignment

The appellate court upheld my claims for intellectual property ownership being granted to students who make discoveries during academic research. The case, known as *Chou v The University of Chicago*, became precedent for students and academic researchers who sought intellectual property rights for their discoveries. (For a copy of the court's ruling, see http://caselaw.findlaw.com/usfederalcircuit/1004756.html).

The appellate court remanded the case back to the district court for plaintiff damages and final disposition.

When, in June 2002, our lawsuit settlement was presented for my signature, the wording of one of its terms disturbed me. Although the document explicitly protected me from any future claims or liability from the suit, it did not protect my husband and my children. In contrast, Roizman himself, his spouse and family, his estate, and his heirs were all named in the released parties (Figure 3A).

My lawyers explained to me that the settlement terms not listing my spouse and children for lawsuit releases, in essence, had little effect in the eye of the law because any retaliatory acts against my spouse and family would be prohibited and judged criminal. I remained uneasy about the terms then. Malcolm was still an associate professor at the University of Chicago and my younger daughter still attended the University High Laboratory School. Could they be at some kind of risk or omen from the lawsuit settlement releases? Was someone at the university planning something that involved Malcolm and my children?

SETTLEMENT AGREEMENT

B. Definition of "Chou Released Parties". For purposes of the releases contained in succeeding section of this Agreement, the term "Chou Released Parties" shall include Dr. Joany Chou, her attorneys, representatives, agents and insurers

E. Definition of "Roizman Released Parties". For purposes of the releases contained in succeeding section of this agreement, the term "Roizman Released Parties" shall include Dr. Bernard Roizman, his spouse and family, estate, heirs, attorneys, representatives, agents, and insurers.

Figure 3A: Terms of Released Parties in the Settlement Agreement

At the same time all of this was happening with the suit, bioterrorism began to take shape at the National Institutes of Health (NIH) and the Department of Defense (DOD). In February 2002, Bernard Roizman was called to chair the blue ribbon panel of scientists to NIH for policy recommendation.

I had not thought about the settlement release until September 2009 when Malcolm died suddenly. Could this be the seed of hatred against Malcolm lurking just beneath the surface? I could not be sure, but I was troubled enough by the university's unfair treatment of me in the lawsuit settlement to wonder whether Malcolm's death was now implicated in it as well.

I panicked at the thought and the connection of the two events; an ill wind had blown my way. I started to pray. *"If the hatred and evil are so destined in our lives, let all of the accusers and debt collectors come to me. Let me burn and let me bear the crucifix indicated by my past. Spare Malcolm and my children from the vengeful gods that traffic in social injustice."* I also prayed that I would see Malcolm again in another world. I wanted to hold his hands and walk beside him, experiencing all that he did. I started to cry as I prayed.

The betrayal by Roizman, my loss of employment, and the legal proceedings took their toll on me professionally and personally as well as on my family. Although Malcolm had been supportive throughout the years of the lawsuit, my inability to pursue my research career in Chicago and my need to get away from the university and the city finally came between us. I wanted to leave and Malcolm wanted to stay because of his tenured position with the university; there was no middle ground, no way to win.

In 2004 we were divorced, and I moved to California to start a new life while giving up everything I had done in science and what I treasured most in the fifteen years of my research career. Although our marriage was over in the next few years, Malcolm and I remained close, tied to the family and to each other until the day that death separated us. We were in constant touch, sending emails to each other on a daily basis, resolving issues and matters of the family, working eagerly on our children's lives and education, including the choice of a medical school that Leigh liked. It was then I began to feel how much I had loved this man and felt guilty that I had not taken care of him, that I had left him to the ill winds that came upon us soon after the lawsuit.

Bernard Roizman was named as a member of the NIAID Blue Ribbon Panel on Bioterrorism and Its Implications for Biomedical Research in the February of 2002. I had no idea at the time that Malcolm would eventually be caught up in the policies that would come from that panel and others like it and the dangers of that area of research. Nor did I realize that Malcolm was already being wounded and disparaged by his colleagues in his own department. Our separated lives made him vulnerable to the changing fortune that would beleaguer him from that day forward.

Malcolm's Changing Fortunes

My successful lawsuit against the University of Chicago and one of its prominent professors was only one factor in Malcolm's changing fortune at the university. Scientists' careers sail or founder on the waters of federal grant money. In the 1990s, the money had begun to dry up for Malcolm, which meant he could not sustain his own research at the university.

Why this lack of funding? The reasons could have been many. Tighter budgets at NIH, from which most of the academic research money came, could have meant fewer grants available. And Malcolm, who had never had that cutthroat competitive edge, could have lost out to others vying for the same awards. Malcolm's field of research—prokaryotic genetics—was also no longer the popular and favored subject of research at the University of Chicago, where many scientists now worked on eukaryotic biology.

Whatever the reasons, the results proved devastating. Lack of funding and the consequent loss of students made Malcolm's research difficult at the laboratory level. Insufficient research activity made it even more difficult to win grant support to breathe life into his career. In a series of events that followed, Malcolm's standing at the university was weakened, compounded by the misfortune of my lawsuit. In 2005, the MGCB Department took Malcolm's laboratory space away from him. That forced Malcolm to take refuge, he hoped temporarily, in Olaf Schneewind's laboratory and left him vulnerable to what was to come.

Academic Fraud Investigation in 2009

The next blow came in early 2009. Although I had left the University of Chicago in 1996 and the lawsuit had been settled in 2002, the university was not done with me. In early 2009 I was told that a committee had been established to investigate a charge of fraud against me. *This is ludicrous*, I thought, when I first learned the supposed basis for the fraud and the person who had made the charge against me.

The accusation concerned a paper I published in 1995 (Figure 3B), fourteen years before, and it had been made by Dale Mobley, a former technician assigned to work on a project under my supervision in 1994. The charge concerned the molecular weight of the putative protein in the paper. For some bizarre reason, Mobley was claiming that I had fabricated the molecular weight of the protein in study and had gone on to make his accusation to other academic professors on other campuses.

Proc. Natl. Acad. Sci. USA
Vol. 92, pp. 10516–10520, November 1995
Biochemistry

Association of a M_r 90,000 phosphoprotein with protein kinase PKR in cells exhibiting enhanced phosphorylation of translation initiation factor eIF-2α and premature shutoff of protein synthesis after infection with $\gamma_1 34.5^-$ mutants of herpes simplex virus 1

(translation initiation/growth arrest and DNA damage proteins/cell stress)

JOANY CHOU*, JANE-JANE CHEN†, MARTIN GROSS‡, AND BERNARD ROIZMAN*

*Marjorie B. Kovler Viral Oncology Laboratories and ‡Department of Pathology, The University of Chicago, 910 East 58th Street, Chicago, IL 60637; and †Harvard–Massachusetts Institute of Technology Division of Health Sciences and Technology, Cambridge, MA 02139

Figure 3B: This paper (1995) and its published content was the subject of fraud by Mobley in 2009.

When I learned of the charges, I was upset but not worried that the investigation would find any truth to the charge. Not only was I fully confident that my published work was accurate to the finest detail, but I also knew that the putative protein under investigation had been molecularly determined and independently verified by several other major laboratories in the United States.

So how had this all come about? As I discovered later, Mobley sent an email to Dr. Keith Moffat, deputy provost for research at the university and Dr. Randy Schekman, a Nobel Laureate of 2013 at Berkeley (Figure 3C). The center of his accusation was about the molecular weight of a Protein I identified at *to be* M_r 90,000 (Figure 3B).

Mobley v Chou

As Mobley began to lay out his tale of fraud claims against me.

> "iv) I noted, on a whim investigation in Dec 2008, having not considered it, that Joanny had actually filled a suit against the professor. The summary "A Victory for the Student" in the Duke bulletin made me think perhaps this aspect was utterly neglected, and I wanted to act civically that a faculty would want to know about it, whether or not it is a major concern of misconduct or not, and whatever the consequence, hopefully to promote the well being of the student and professor, and again toward the well-being of the profession and university." (Figure 3C)

And specifically the charges were as follows:

> "My charge was what I brought to the professor in 1998, i) that Dr. Chou manipulated data to get a false result, ii) that it was brought to her and the professor's attention, iii) that they pronounced it faulty, iv) that she failed to reproduced it, though ignored the failure, v) that after I left I discovered it nonetheless published, when visiting the library of another university" (Emphasis added, Figure 3D).

Moffat quickly raised the issue with the university legal counsel, Ted Stamatakos, in an email, asking if Mobley's claim should be considered "a serious credible accusation of academic fraud. (Figure 3E)

> "Ted-
> Does this e-mail constitute a serious credible accusation of academic fraud? How do I/we proceed if the answer is "yes"? If "no"?
>
> Note that the editor of PNAS, Randy Schekman, is copied so this is not simply a matter internal to the UofC."(Emphasis added, Figure 3E)

They decided that it was. Moffat then sent the e-mail below to initiate a series of fraud investigation within the university on one of my papers in1995 by Chou, Chen, Gross and Roizman. (Figure 3B)

> I'm the individual in the Provost's Office responsible for overseeing allegations of academic fraud. Here's some background info for our meeting tomorrow at 1.00 in CLSC1106.
>
> In response to my top e-mail below, Ted Stamatakos from Counsel's Office advised that we must conduct an "initial enquiry" according to the University's policies on academic fraud. These policies are to be found at ... Ted and I have discussed this and earlier e-mails involving Mobley and Schekman (copied at the foot of the e-mail chain) at length.
>
> The article in question appears to be Chou, Chen, Gross and Roizman, PNAS 92, 10516-20 (1995): and the figure questioned appears to be Fig 4 (not Fig 3 as stated by Mobley). (Figure 3B)

In March and April of 2009, just months before Malcolm's death, the university initiated a formal fraud investigation against me while Malcolm stood on the sideline watching in horror. He knew that a fraud investigation of a scientist could undermine his or her entire scientific integrity, career, and reputation. Even if unproved, the investigation itself could throw a shadow over the scientist's reputation and credibility.

In April 2009, Anthony Mahowald, former chair of the MGCB department, finally called me on the phone to testify about my work.

I said calmly, "Tony, did you talk to Bin He about this figure?"

Tony sounding pleased that he had already questioned Bin He on this point) replied, "Yes, I talked to Bin He, and he did not know if the figure warranted fraud. He simply was not familiar with your work."

Bin He was a postdoctoral student who had come to Roizman's laboratory a year before my departure and been assigned to me to mentor. Over the following year, I taught him everything about the project on which I published the paper, a project I had developed in Roizman's laboratory from beginning to end. I couldn't understand his professed ignorance about the figure, which not only was the basis of the project I had worked on but actually appeared in a patent that Roizman and He were later granted. That He was one of the supposed inventors of the process in the patent meant that Roizman and He, and the University of Chicago, were claiming that He had participated in the work and understood everything in it.

"Tony, have you seen the patent by Bernard Roizman and Bin He?" I asked. "This figure that you are investigating was in Roizman and He's patent. And He now claims he doesn't know the accuracy of this figure?"

Mahowald paused, clearly unfamiliar with the patent and not sure how to continue; I had evidently derailed the course he had planned for the conversation. When he continued, he sounded annoyed. I wasn't playing my part of a conjured researcher as he expected.

"I am not talking about patents in lawsuits that you had with the University. I am talking about this figure, this figure in your 1995 paper that has been called fraud and under investigation by the University Fraud Committee. I am here to talk about the fraud charges brought by Dale Mobley about this paper on which you were the first author."

I sighed, knowing I wasn't going to get my point across to him in this conversation. "Tony, do you have a pen with you?"

"Yeah," Tony said impatiently.

"Write this number down. Patent #5,795,713 published by inventors Roizman and Bin He."

Tony clearly did not understand my point and the problems this investigation could create for the University of Chicago itself. First, the fraud investigation into a figure that had been patented by UC scientists could destroy the validity of the university's patent and their academic reputation. Second, it was fraud to have taken figures and part of my work to use in a patent while listing Roizman and Bin He as inventors when Bin He had not participated in the original work and was now claiming he could not certify the figure's accuracy.

I had been preparing to send Tony copies of my notebooks documenting the experiment (since the university was claiming it couldn't find the originals of the work supporting this figure), but it proved unnecessary. Not surprisingly (to me), soon after Tony and I talked on the phone, the investigation was closed and the committee, which had found no wrongdoings on my part, was disbanded. In fact Tony later commented that the lawyers who represented the University of Chicago in my lawsuit told the investigators not to get into that mess again. They closed the case.

All of it left me feeling angry and wondering what the motive for the whole farce had been. After all, Mobley himself had said in his original email that he had only looked into the matter of the figure in my research "on a whim." And what scientific standing did Mobley have for doing so? He held only a bachelor's degree and had gone on to become a seminary student after 1995.

Furthermore, although I was the first author on the paper in question, I wasn't the only author. The work was jointly carried out by and coauthored with professors Dr. Jane-Jane Chen, from Harvard-MIT, Division of Health Sciences and Technology,

Cambridge, Massachusetts, and Dr. Marty Gross, a faculty member in Department of Pathology at the University of Chicago, whom I consulted and shared expertise on this subject. So why, if the university was really investigating a legitimate accusation of fraud, was I the only one under investigation?

None of it made sense unless the actions were perceived as punitive, a punishment for my earlier lawsuit. The time of the fraud investigation was just a few months before Malcolm's death. I wasn't sure what Mobley had hoped to gain by the whole affair, unless he somehow thought he had been harmed by my lawsuit.

For various players at the University of Chicago, perhaps they had seen it as a chance to punish me and further isolate Malcolm. I didn't know, but though I had believed absolutely in the accuracy of my work, the investigation itself had taken its toll, leaving me mentally, emotionally, and physically depleted.

iv) I noted, on a whim investigation in Dec 2008, having not considered it, that Joanny had actually filed a suit against the professor. The summary "A Victory for the Student" in the Duke bulletin made me think perhaps this aspect was utterly neglected, and I wanted to act civically that a faculty would want to know about it, whether or not it is a major concern of misconduct or not, and whatever the consequence, hopefully to promote the well being of the student and professor, and again toward the well-being of the profession and university.

Sincerely,

Dale Mobley
dale.o.mobley@gmail.com
PO Box 1581
Elizabeth City, NC 27906
(252) 335-5890

Figure 3C: e-mail from Mobley with fraud charges against Chou

****

My charge was what I brought to the professor in 1998, i) that Dr. Chou manipulated data to get a false result, ii) that it was brought to her and the professor's attention, iii) that they pronounced it faulty, iv) that she failed to reproduce it, though ignored the failure, v) that after I left I discovered it nonetheless published, when visiting the library of another university.

Figure 3D: Mobley laid out fraud charges to the university

****

From: Keith Moffat
Sent: Mon 2/16/2009 9:54 AM
To: stamatak@uchicago.edu
Subject: FW: Chu/Roizman/PNAS Publication

Ted -
Does this e-mail constitute a serious, credible accusation of academic fraud? How do I/we proceed if the answer is "yes"? If "no"?
Note that the editor of PNAS, Randy Schekman, is copied so this is not simply a matter internal to the UofC.

Figure 3E: Moffat seeking advice from UofC legal Counsel over fraud charges

****

From: "Keith Moffat" <moffat@cars.uchicago.edu>
Date: February 16, 2009 3:56:43 PM CST
To: <rfehon@uchicago.edu>
Cc: <stamatak@uchicago.edu>
Subject: PNAS Publication CONFIDENTIAL

Rick -
I'm the individual in the Provost's Office responsible for overseeing allegations of academic fraud. Here's some background info for our meeting tomorrow at 1.00 in CLSC1106.
In response to my top e-mail below, Ted Stamatakos from Counsel's Office advised that we must conduct an "initial enquiry" according to the University's policies on academic fraud. These policies are to be found at http://old.uchicago.edu/docs/policies/provostoffice/fraudpol.pdf . Ted and I have discussed this and earlier e-mails involving Mobley and Schekman (copied at the foot of the e-mail chain) at length.
The article in question appears to be Chou, Chen, Gross and Roizman, PNAS 92, 10516-20 (1995); and the figure questioned appears to be Fig 4A (not Fig 3 as stated by Mobley).
Best,
Keith

Figure 3F: Moffat initiated a university wide fraud investigation against Chou over fraud accusation brought by Mobley

July Call and Nightmare

A curious incident also happened at this time, though I was not aware of it until later. Malcolm was given his first plague vaccine product, in May 2009, and fell sick. No one connected it to the vaccine, but fortunately he was saved because doctors gave him antibiotics to treat a lacerated toe at the same time.

On July 15, 2009, I called Malcolm late in the evening. I wanted to update him on the fraud investigation, which had closed, and urge him to leave Chicago and come to California. It was something I had urged before, but after the most recent events, I feared that his situation had only worsened. I had no idea, however, just how badly things were going for him.

"Mal," I said affectionately, when he picked up the phone.

Before I could get out another word, he began shouting, his voice filled with anger and something even more like terrifying fear. This wasn't like him. Malcolm had always been a soft-spoken and gentle man; he did not ever yell.

I could not understand most of what he was saying, but a few words stood out: "Dangerous . . . Refuse . . . Deadly!" I felt panicked. What was wrong with Malcolm? What had set him off to be screaming in a manner so uncharacteristic of him? Why was he so angry or fearful?

I listened for the better part of three minutes while Malcolm shouted and continued his rant. Only when he stopped for a moment could I break into the conversation. "Malcolm! Malcolm! Calm down."

Thinking it might be the fraud investigation that had disturbed him and led to this uncharacteristic breakdown, I said quickly, "Listen, I need to tell you something about the academic fraud."

Whether that was the source of his trouble, my words worked. He calmed down and listened.

"There was no fraud, Malcolm!" I told him emphatically. "Two other major laboratories already proven my point. They had cloned the DNA sequence encoding this putative p90 protein that interacts with PKR like I had demonstrated in this paper. The molecular weight was 90kdal right on the dot."

I continued, "How could the university investigate fraud of a figure in my paper when that figure had been assigned to Roizman and He's patent. The fraud investigation would have rocked and destabilized the university's patent; that is the last thing the university would want to do. It was nothing but a witch hunt."

There was more silence on his part, but I knew he heard me. I realized that, whatever else troubling Malcolm, he had been afraid for me, afraid that the university might prove something against me. Perhaps someone there had even been telling him lies about my work.

Malcolm finally asked me to send him the paperwork I had received on the close of the investigation. I promised to send the paperwork that he wanted but never got the chance to send it before his mysterious death just weeks later.

I realized, from our conversation, that Malcolm had been severely affected mentally and emotionally by the fraud investigation and privately agonized over the proceedings that could destroy my reputation as a scientist as well as his. After all, Malcolm knew me and watched me grow from a college student technician to a professional scientist by the time I left Chicago. Malcolm had such a gentle soul that he would never doubt my work and my integrity in science on his own.

We ended the conversation for the night. My mind was exhausted and my body weak from the strain of the bogus investigation, which came on top of all the stress I had experienced years before with my lawsuit. When I lay down to sleep that night and the next, I just wanted relief from everything. I wouldn't get it.

Nightmare or Premonition?

The day after my conversation with Malcolm on July 15, I had the most vivid dream I've ever had. In the dream, I walked into a bright white room that was crowded with people I did not know. I had to squint just to see anything because the light was so bright. There was little in the way of furniture in the room, and as I moved about, I noticed that the people were all leaving. The room was almost empty when I heard a voice from some place behind me.

"Malcolm is in the attic," the male voice said as I struggled to see who was speaking. I looked up to the attic and could see the edge of the attic next to the ceiling. I struggled to see Malcolm. I walked about that room searching for a staircase I could take to the attic but I could not find one. The voice said that Malcolm was at the deep end of the attic.

Then came a different voice spoke, like the voice of a woman or a small child. "He was put in the attic by people in his department with just enough food and water," the voice explained.

My mind struggled to understand what was being said to me. Was it his choice to be in the attic or did it happen against his will? Was he chained to the attic, and if so, why could he not unchain himself and get out? I wondered how much food and water he had left and for how long. Would Malcolm return to us from the attic after the food was gone, or would that be the end of his life? I walked about the room to find an angle where I could see into the attic and locate Malcolm. I stared hard at where I thought the attic was, trying to see any signs of Malcolm. I wanted to take his hand

and bring him down, but there were no way to the attic. I thought he must be at the furthest ends of the attic, not visible from where I was.

Then I heard the soft-spoken voice of Malcolm murmuring some words that I could not comprehend. His voice was weak. It was then that I felt myself growing impatient and even angry. Why had Malcolm been put in an attic and left alone? Was he being punished? If he was, what was he being punished for?

Out of nowhere, I was suddenly aware that my daughters were standing next to me in the room. I grabbed their hands and asked them to find me a ladder so I could climb up to the attic to seek Malcolm. I asked them if they knew why he was put there. They did not respond, so I frantically searched the white room for any type of ladder. There was none. A sudden panic gripped my soul, and I began to sweat profusely. I was just about to scream when my eyes opened and I realized it was just a dream.

Now two months later, as my plane prepared to land in Chicago, I wondered if that dream had been a premonition of what was to come. An eerie feeling crept over me.

4

U.S. BIOTERRORISM SLOWLY CREPT INTO OUR LIVES

Abruptly, my thoughts shifted like winds in a storm to something horrid and disturbing. Thoughts of bioterrorism in the U.S. and the biological warfare development slowly crept into me, permeating everything with horrific possibilities.

U.S. Bioterrorism

In February 2002, soon after the 9/11 attacks that shook the nation, the George W. Bush Administration ordered the Department of Defense (DOD) and NIH to set goals to defend against bioterrorism. The goal of the program was to study biological and chemical weapons in biosecurity and biodefense resource (Figure 3A). It was a huge program supported by billions of dollars in federal grants to the national labs that NIH had selected and designated (Figure 3B).

Bioterrorism, Biodefense Research

http://www.genomenewsnetwork.org/articles/10_03/biodefense.shtml

NIH Distributes Biodefense Funds on October 17, 2003

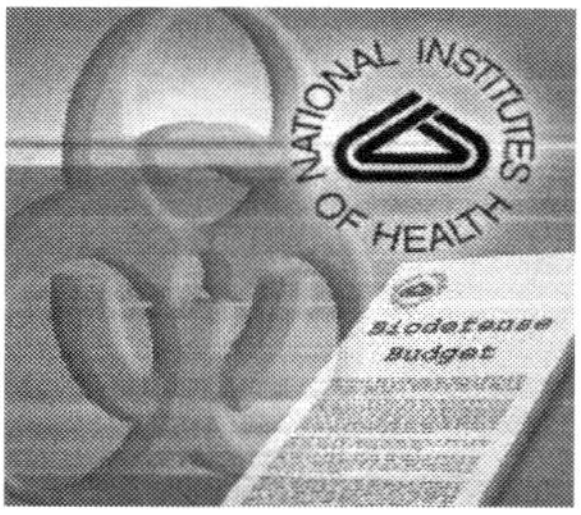

In response to increasing concern about the nation's vulnerability to attack with biological agents, the U.S. government is allocating generous sums to biodefense research in six major areas, including genomics, diagnostics, and vaccine development.

The President's 2003 budget allocated $1.55 billion to the National Institutes of Health (NIH) for biodefense research. The lion's shareó$1.16 billion went to the National Institute of Allergy and Infectious Diseases (NIAID), the principal NIH institute that supports biodefense research. The rest was divided among the other institutes.

Awards include a total of $350 million to establish nine Regional Centers of Excellence for Biodefense and Emerging Infectious Diseases Research. The purpose of the centers is to conduct research aimed at understanding the biology of pathogens such as anthrax, plague, Ebola, and smallpox, and to develop new vaccines, antibiotics, and other therapies for preventing disease.

The NIAID awarded another $85 million to five Cooperative Centers for Translational Research on Human Immunology and Biodefense. These centers will focus on the human immune system and its response to bioterror agents.

And just before the fiscal year ended on September 30, NIAID announced awards to construct and support eight regional and two national biocontainment laboratories *throughout the United States.*

Each national laboratory will receive $120 million, while regional facilities will receive between $7 million and $21 million. Much of the research funded by the Regional Centers of Excellence will be conducted in these biocontainment facilities.

NIAID has also begun awarding contracts to companies to develop vaccines as part of their Biodefense Partnership program. A total of 31 grants have been awarded to pharmaceutical and biotechnology companies to develop products for biodefense, including vaccines, antibiotics, and diagnostic tests.

Figure 4A: Federal funds distributed to bioterrorism, Biodefense Research. (Emphasis added)

Regional Center of Excellence (RCE) National Labs

(http://www.niaid.nih.gov/LabsAndResources/resources/rce/Pages/sites.aspx. (Figure 4B).

Ricketts Regional Biocontainment Laboratory

A group of research laboratories at the University of Chicago were chosen to become the "Center of Excellence of the Great Lakes Region" (GLRCE) (Figure 4B top). Olaf Schneewind was the principle investigator of the GLRCE national lab (Figure 4B, bottom) and Keith Moffat was the principle investigator to head the Howard T. Ricketts Regional Biocontainment Laboratory (Figure 4C). The consortium included twenty-seven neighboring research institutions in the Great Lakes region with a mission to foster and sustain productive research environment in which academic divisions overlapped in their interests and thrived on collaboration. This was where it all began.

Regional Centers of Excellence for Biodefense and Emerging Infectious Diseases (RCEs)

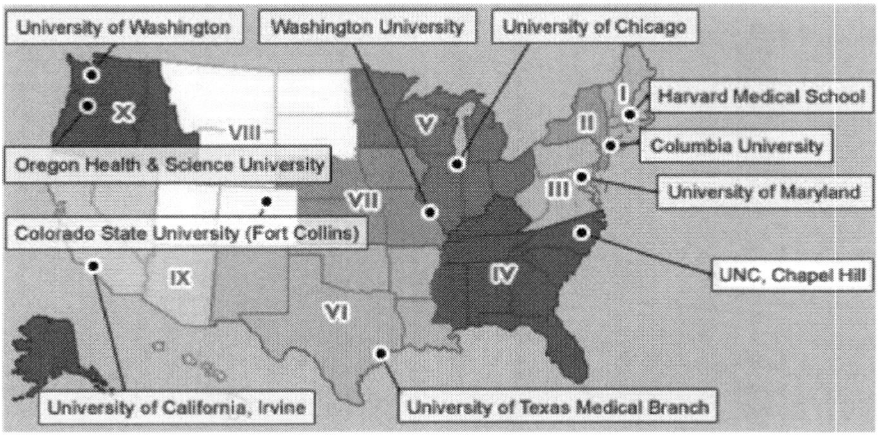

Region I
Harvard Medical School
New England Regional Center of Excellence
For Biodefense and Emerging Infectious Diseases
PI: Dr. Dennis Kasper

Region II
Columbia University
Northeast Biodefense Center
PI: Dr. W. Ian Lipkin

Region III
University of Maryland, Baltimore
Mid-Atlantic Regional Center of Excellence
For Biodefense and Emerging Infectious Diseases
PI: Dr. Myron Levine

Region IV
University of North Carolina, Chapel Hill
Southeast Regional Center of Excellence for
Biodefense and Emerging Infectious Diseases
PI: Dr. Fred Sparling

Region V
University of Chicago
Great Lakes Regional Center of Excellence
For Biodefense and Emerging Infectious Diseases
PI: Dr. Olaf Schneewind

Region VI
University of Texas Medical Branch
Western Regional Center of Excellence
For Biodefense and Emerging Infectious Diseases
PI: Dr. David Walker

Region VII
Washington University
Midwest Regional Center of Excellence
For Biodefense and Emerging Infectious Diseases
PI: Dr. Herbert "Skip" Virgin

Region VIII
Colorado State University (Fort Collins)
Rocky Mountain Regional Center of Excellence
PI: Dr. John Belisle

Region IX
University of California, Irving
Pacific-Southwest Regional Center of Excellence
PI: Dr. Alan G Garbour

Region X
Oregon Health & Science University
Pacific Northwest Regional Center of Excellence
PI: Dr. Jay Nelson

Figure 4B top: Ten National Labs selected by NIH to engage in bioterrorism research. Bottom: Principle Investigator of the National Laboratories listed

- http://www.fas.org/biosecurity/resource/map/ucrbl.htm

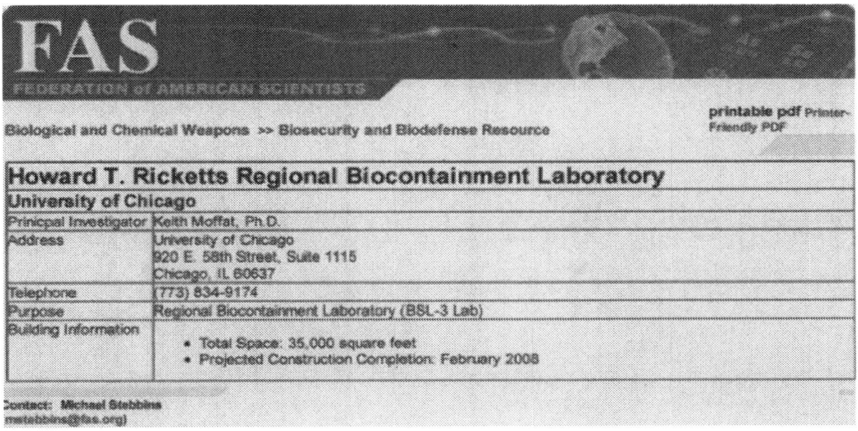
FAS
FEDERATION of AMERICAN SCIENTISTS

Biological and Chemical Weapons >> Biosecurity and Biodefense Resource

printable pdf Printer-Friendly PDF

Howard T. Ricketts Regional Biocontainment Laboratory	
University of Chicago	
Principal Investigator	Keith Moffat, Ph.D.
Address	University of Chicago 920 E. 58th Street, Suite 1115 Chicago, IL 60637
Telephone	(773) 834-9174
Purpose	Regional Biocontainment Laboratory (BSL-3 Lab)
Building Information	• Total Space: 35,000 square feet • Projected Construction Completion: February 2008

Contact: Michael Stebbins
mstebbins@fas.org)

Figure 4C: Moffat was the PI of the Howard T. Ricketts Regional Bio-containment lab

Bioterrorism Research Development

Francis Boyle, professor of international law at the University of Illinois College of Law, had written in his book *Biowarfare and Terrorism* (Clarity Press, 2005) that "[t]he United States has had an extremely aggressive, offensive biological warfare program dating back to World War II." He went on to state:

> The NeoCons [neoconservatives] in the Bush Jr. administration have tried to sell this biowarfare arms race to the American people and the U.S. Congress as well as to the rest of the world by disingenuously claiming that it is all for the purpose of "defense," which, as previously explained, is not a permissible exception to the Biological Weapons Convention in the first place . . . The massive proliferation of biowarfare technology, facilities, as well as trained scientists and technicians all over the United States courtesy of the NeoCon Bush Jr. administration will render a catastrophic biowarfare or bioterrorist incident or accident a statistical certainty. [Frank A. Boyle, *Biowarfare and Terrorism*, Clarity Press, 2005]

Boyle's book, as a review explained, makes a case for "how and why the United States government initiated, sustained, and then dramatically expanded an illegal biological arms race with potentially catastrophic consequences for the human species." Another explanation of bioterrorism, which has been cited time and again by various sources, states:

> Biological weapons may be employed in various ways to gain a strategic or tactical advantage over an adversary, either by threats or by actual deployment . . . These agents may be lethal or non-lethal, and may be targeted against a single individual, a group of people, or even an entire population. They may be developed, acquired, stockpiled or deployed by nation states or by non-national groups. In the latter case, or if a nation-state uses it clandestinely, it may also be considered bioterrorism. [*Microbiology*, Boundless, 2013, p. 1306]

Plague Vaccines: Live Attenuated Vaccine

The project at the University of Chicago involved the development of "plague vaccines" using a class of biological molecules known as the live attenuated vaccine strains of plague. The work was carried out in the laboratory of Dr. Olaf Schneewind, the principle investigator of the Great Lakes Regional Center of Excellence National Lab, professor and chair of the University of Chicago Microbiology Department.

I hold a Ph.D. in virology from the University of Chicago and have published numerous articles and acquired several patents in the biology of the herpes simplex virus. My first patent revealed the nature of a live attenuated vaccine of herpes simplex virus that I had created and sought for its vaccine property. The live attenuated vaccine strain in general was a strain of infectious agent that had the ability to replicate itself in the host of animals. Because of the attenuation, either engineered in the laboratory or obtained in the wild, the strain maintained a state of "reduced" or "attenuated" virulence relative to that of the wild type. While multiplying, the strain would elicit immune response in the host animal. Pathogens with these properties had been classified as live attenuated vaccine strain to protect humans against infection (Figure 3E).

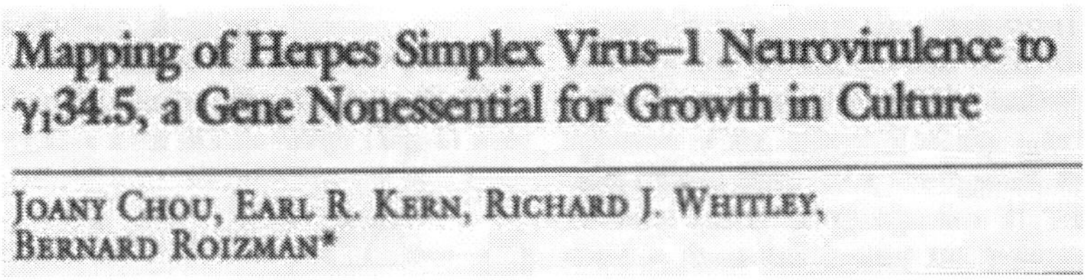
Mapping of Herpes Simplex Virus–1 Neurovirulence to $\gamma_1$34.5, a Gene Nonessential for Growth in Culture

JOANY CHOU, EARL R. KERN, RICHARD J. WHITLEY, BERNARD ROIZMAN*

Figure 4D: Herpes Simplex Virus-1, $\gamma_1$34.5 deletion mutant, is a live attenuated vaccine strain

For purposes of U.S. bioterrorism research at Chicago, KIM D27 had been designated as a live attenuated vaccine strain of plague. Plague bacteria is a blood-borne pathogen, toxins released would infiltrate bodily organs and cause severe headaches and body aches and a massive organ failure. Death ensued in individuals inflicted with septicemic plague at 100 percent fatality without antibiotic intervention (KIM D27 Agent Profile).

Despite all mysteries surrounding Malcolm's sudden death, I repeatedly asked the same question, "How could this happen at the national laboratory? And Malcolm, a molecular biology expert?"

Institutional Biosafety Committee (IBC)

The federal bioterrorism program required that each national lab observe biosafety, a biosafety net set forth by a university-based Institutional Biosafety Committee (IBC). The purpose of the IBC was to work with NIH Office of Biotechnology Activities (OBA) and principle investigators of the university to formulate a set of biosafety protocols to enforce biosafety standards to the working laboratories at the university. A sample of IBC minutes on 6/20/03 is shown in Figure 4E.

In addition, national laboratories engaging in biodefense and bioterrorism research were required by NIH policy to submit "Agent Profiles Form" and "Protocol Submission" for the bioterrorism project on a semi-annual basis for NIH to review. The Agent Profile Form and the Protocol Submissions had to be approved by the university IBC followed by Select Agent–IBC headed by Dr. David Pitrak at the university before submission to NIH for review.

Malcolm was, by anyone's standards, an experienced, reputable scientist and an expert in microbiology. In fact, there was no one like him. In addition to class teaching at the undergraduate and graduate level, Malcolm had served on the university IBC as a voting member for more than fifteen years. He attended monthly biosafety meetings willingly and diligently and had a record of perfect attendance. More remarkably, Malcolm with fifteen faculty members of the IBC drafted the Biosafety Standard Protocol that was in practice by all biological laboratories on campus and at the national laboratory. Because of his unique experience and knowledge in biosafety, there was not the remotest possibility that he could have contaminated himself and sealed his fate. There had to be more to the mystery.

My thoughts continued to wander. *The university had on its staff world-renowned microbiologists, plus staff and students whom Malcolm once taught and many more of his past students who held important positions in the current administration and in private sectors. In addition, the university's being a national lab, the Center of Excellence, surely would take good care of the lives of their faculties, more significantly, the reputation of the university represented by the faculties who wore academic robes and hoods,* I thought.

I quickly tossed out the unpleasant thoughts that rambled through my mind, but traces of them still lingered. I hoped and prayed that Malcolm's death was not related to plague, but the scientist in me had more to ask: *Was it possible . . .? Could it be possible . . .?* I was convinced that there had to be more to the story. I needed more time and more information to confront the mystery, the ill winds that hovered over Malcolm.

I took a deep cleansing breath as I thought about how hard the next few days were going to be for my family and me. What I did not realize at the time was that it was not going to be just a few tough days. It was going to be a tough few years with more questions to be asked and more answers to be sought after.

I suddenly remembered that on July 15 in our conversation, Malcolm told me that he begged—and he most definitely used the word "begged"—them at the departmental faculty meetings but to no avail and now he was dead. *Was Malcolm the scapegoat for the lawsuit I filed against the university and won and the fraud investigation on my science integrity? What issues were at stake that caused him to beg his colleagues in July 2009? Begged for his life? Did his death seem imminent to him and to others in his department? Clearly he had suffered a mortal blow; perhaps a more significant issue had been at stake? Perhaps he was concerned over the safety of the bioterrorism products that were to be used for human experimentation on laboratory personnel? Malcolm had been a biosafety officer in the university IBC for more than fifteen years! What about the stupid lies claiming Malcolm was not the father of my children? How could anyone deal with all these complicated issues? I suddenly had a horrid notion that perhaps his death might not be an accident but a preplanned, premeditated gang-style execution.*

What was wrong with the University of Chicago? How could they humiliate one of their professors the way they had Malcolm? It was beneath the conduct expected of faculty members who wore the university robes and hoods. What had they done to Malcolm who had always been the gentle soul and spirit that Beckwith and others had come to know? What did Malcolm do to deserve this treatment?

I remembered that Malcolm once told me, "I went to see Moffat, the academic dean who oversees faculty appointments."

"Yeah? What did he say?" I responded.

"He helped me."

"Really?" Waiting for him to say more.

"Moffat assured me that no one in the department would touch me or take my tenure away," Malcolm added. "Moffat is the one who oversees faculty appointments in the Provost's Office. He knew everything. I believe he is the one man in the department who can right all wrongs. He told me not to worry. Moffat is confident that nothing like this would slip by him. He is a powerful man, and he is helping me." Malcolm rushed through these rambling words with such faith that no one could contradict him.

Malcolm went on rambling about two things Moffat had told him: "Don't worry about the people in the MGCB and the Microbiology Department. Nothing is going to happen to your tenure without my approval." Malcolm continued, "He also told me not to put anything in writing, in emails. Don't tell anyone about this even in phone conversation. People are listening and reading your emails."

Apparently, Moffat was very specific and concerned about this. "JC," Malcolm told me afterward, "don't write me any emails about anything we have talked about. It's not safe. Someone could be watching and listening."

After getting this directive from Moffat, Malcolm never once entered anything else into his office computer. But I never really understood what subject matter was involved, what it was that Moffat specifically did not want Malcolm to talk about.

Was it the bioterrorism project protocol that Moffat had wanted Malcolm to be silent about? Had Malcolm been the target of bioterrorism warfare under development? I was horrified by this thought. I only knew that Malcolm spoke so highly about Moffat and his soul-saving power that I had no reason to question Moffat.

But Moffat had not saved Malcolm, despite his encouraging words, had not protected him from the evil winds clearly present in the university. Again I recalled Malcolm's ranting. These words haunted me because they were so unlike Malcolm. Something had set his world in a tailspin. "DANGEROUS . . . REFUSE . . . DEADLY." He had tried to tell me something but had not gotten his meaning across. *What was so dangerous?* Who refused, and what was it that Malcolm felt was so deadly?

I later learned that Moffat was the principal Investigator of the Howard T. Ricketts Regional Biocontainment Laboratory, a job in the central administration of U.S. bioterrorism at the University of Chicago (Figure 4C).

I sank in my plane seat, allowing my thoughts to drift. Soon I found myself in a masquerade ball. There was music and dances, jokers and clowns, but everyone wore masks. I could not find Malcolm anywhere. I looked for him in the crowd. I crawled from underneath one table to another and found no trace of him. Maybe he was wearing a mask as well. There were more laughter and sneers, and the music played louder. I sank to the floor, and people began to step on me.

These were terrifying moments that I had to endure privately on this long journey to Chicago. I knew something was unnatural in Malcolm's death, but above all else I had to be cautious.

When the plane finally landed and the cabin lights came on, I woke up in a sweat. I found myself in the midst of passengers eagerly waiting to get off the plane. All the masks and laughter were gone. My daughter was sitting two seats from me, buried in her own grief, paying no attention to me.

My Silent Promise and Prayers

Moments before I deplaned, I made an earnest silent promise to Malcolm that I would stop at nothing to search and uncover the truth about what happened to him. I would not let anyone or anything stand in my way. I suddenly felt the power. I cried silently to myself, *"Malcolm . . . Malcolm, here I come. I am taking you home."*

I would learn years later while sorting through some documents that Schneewind and Bill Blaylock attempted to get into Malcolm's office computer soon after his death. They logged in using Blaylock's password. Once they got into his computer, they discovered that Malcolm had not made an entry since June 26, 2009. I would put the pieces together and realized that June 26, 2009, two and half months before his death, must have been the day when Malcolm went in to talk to Moffat about his life-threatening fears. Moffat told him not to worry and not to enter anything in his computer. Clearly Moffat knew something that Malcolm did not, and Malcolm never told anyone about the bioterrorism warfare program at the university, not even family members.

Clearly something diabolical had been going on, and Malcolm had been manipulated, deceived, and riddled from early on. Had there been a game of chess at the university, Malcolm certainly was one of the pawns.

Our plane landed at O'Hare airport in Chicago at precisely 5:00 a.m. on September 14, 2009. I felt a crisp blast of fresh air brushing over my face as Brooke and I stepped on the tarmac. My heart was heavily laden with sorrow.

I hardly noticed anything new or different about the city I had lived in for twenty-four years. It was the city I left behind in 2004. I simply wanted to pick up Malcolm and help him on his last journey home.

Never at any moment of our lives together had we imagined that Malcolm would exit the university on a stretcher, after painfully and horribly enduring agony that evil forces imposed on him. I had yet to reconcile that his smiles and laughter were gone, that they lay under the blanket of death that covered him on this day.

Brooke and I were both exhausted and our souls ached with grief, but I knew that this was only the beginning of all sorrow. All realities of death and despair would completely consume us in time. Now I needed to pace myself and be strong for my daughters, to contend with any ill forces that would come our way. I thought about my strange dream again and wondered why I had it and if it meant anything. I had the awful feeling that the dream had been a warning, clearly telling me that something diabolical was going on in the final days of Malcolm's life. *Had Malcolm been a victim whom they tried to keep him silent about something. Now he was dead. What was going on?*

We were greeted at the airport by some old friends who took Brooke and me to Malcolm's home on the south side of Chicago. It was the same house where Malcolm and I had lived for twenty-three years from 1981, a year after we had moved to Chicago. The home bordered the north side of the University of Chicago campus. The sight of it brought back a rush of memories.

Within an hour, my other daughter, Leigh, flew in from Boston where she was a student at MIT, Malcolm's first alma mater. Malcolm's mother, Dolores, also arrived from Washington, D.C.. Not long afterwards, Malcolm's father arrived from New Orleans. Everyone was in shock over Malcolm's death, and no one could understand what had happened.

The University of Chicago

Institutional Biosafety Committee

5841 S. Maryland Avenue
AMB S-152 ♦ MC 1108
Chicago, IL 60637

Corrected Minutes of June 20, 2003
3:00 PM – [redacted]

In attendance:

Voting Members		**Ex-Officio Members**
Kenneth Thompson	James Mastrianni (left prior to vote)	Steve Beaudoin
Richard Hiipakka	Louis Philipson	
Malcolm Casadaban ←	Mary Ellen Sheridan	
Helena Mauceri	Craig Wardrip	
Rima McLoed	Gopal Thinakaran	

Guest	**Staff**
Claude Bake, Hospital Safety	Bill Pugh

Absent:

Voting Members	**Ex-Officio Members**	**Staff**
Walter Stadler	Manfred Ruddat	Pamela Postlethwait
George Daskal	Russ Herron	
Clara Gartner	Michael Holzhueter	
	Steve Seps	
	Markus Schaufele	
	David Pitrak	

I. Protocol Review:

The following protocols were reviewed at the meeting, with the disposition noted:

PR# Category/Investigator/***Disposition***

708Ad 03 Amendment/Schneewind, Olaf/***Pending-Conditions*** (7-1)
This amendment requests to add the subcutaneous infection of A/J mice with spore preparations of *Bacillus anthracis* strain Sterne. After an injection of 0.1ml of spore suspension in PBS (dilution of spores 1 x 10^4 or 1 x 10^6 cfu) in the flank, the mice will be observed in 8-hour intervals for acute lethal disease development

for a maximum of 7 days. All injected animals will eventually be euthanized, either after the development of acute lethal disease or after the 7 days. After sacrifice, the liver, spleen, lung and heart will be removed. All tissue samples will be placed in sterile plastic bags, weighed and homogenized. The bacilli will then be removed from the tissue suspension and counted by dilution and colony formation on BHI agar. The number of bacilli in the four organ systems will be recorded, along with the time to development of acute disease for each animal. While this protocol is currently listed as a BL2 protocol, the existing approved animal work with rabbits is only ABSL1, so this amendment changes the maximum ABSL from 1 to 2.

The committee members initially questioned the need to review and vote on this amendment at this point in time, since it was just submitted on June 11th. It was explained that during review of an NIH grant, one of the reviewers recommended that a few pilot studies be performed. The amendment covers the initial pilot study and the anticipated future studies.

The committee also discussed at great length the possibility that the Sterne strain of anthrax is more virulent for some individuals than previously indicated since this particular mouse strain exhibits such great susceptibility to this "non-virulent" agent. After a great deal of discussion, the committee determined that since this was the strain that was used during vaccinations, many people have been exposed to this agent during the vaccination process, and that the CDC considers this to be a BL2 agent, this is properly classified as BL2. One member did disagree with the majority and felt that since there is a mouse strain that is highly susceptible, there could be individuals with the same degree of susceptibility. This member had concerns about the pathogenicity if large amounts of this material was either ingested or inhaled. Also, this member questioned the possibility of lateral gene transfer that would result in the Sterne strain becoming more virulent. This member wanted the investigator to provide referenced data to address these issues. Since this member did not receive support for these viewpoints and thus, it was determined by the committee to not ask these questions of the investigator, this member voted against the motion to approve of this amendment with the conditions as noted below.

Some members asked about the possibility that the Sterne strain could be mixed up with the Ames strain. It was noted by other members of the committee that since the two different strains would not be used in the same facilities this was not a possibility. In its discussion of the virulence of the A/J mouse strain to the Sterne strain, the committee did question if this particular strain also shed the spores. Since the shedding of spores would impact the decontamination procedures of the caging materials and increase the likelihood of an exposure while handling the animals, the committee did decide to require clarification on that issue. Following extensive discussion of this amendment, the committee voted 7-1 to approve the amendment pending satisfactory answers to the following questions:

'0/2003 IBC Meeting Page 2 of 3

Figure 4E: Minutes of IBC meeting on 6/20/03. Note attendance list

5

LAST VISITATION WITH MALCOLM

On September 14, 2009, at 5 a.m., when we arrived in Chicago, I wanted more than anything to see Malcolm. Emotionally, I had not yet accepted the fact that Malcolm had already left us, and I felt like running desperately in all directions to look for him. A dense fog blinded me, and I struggled to keep running until I became exhausted and breathed heavily. Yet there was no sight of him.

The cold reality of death began to seep in as we approached the shadowed walls of the morgue. My daughters walked on either side of me, a hand of each tightly clasped in mine. We held on tightly, supporting each other as we took this the dreadful journey. Malcolm's mother, Dolores, dragged her feet next to us. Together we wanted to see Malcolm.

My heart was beating wildly. Mixed with the grief and bewilderment was a sense of guilt. I wanted to see Malcolm and ask for his forgiveness. I felt I had not taken care of this man, the man I had been married to for almost three decades, the father of my two children and my dear friend. I had left him behind to this awful fate after our divorce in 2004. The fact that I couldn't have foreseen this ending didn't matter; I wanted to ask for his forgiveness.

Questions and anxiety flooded me, some inane, some heart wrenching, tumbling over one another in my mind. *Was Malcolm experiencing the cold temperature of the morgue? Did he utter any words during his last hours of struggle? Was he of clear mind when he talked to the doctors before he died? Did he ask to speak to his loved ones while he lay dying? Was he alone, all alone, when he took his final breath?* With each question came more tears.

It was nearly noon, eighteen hours after Malcolm had passed. We were brought to the viewing room; reluctantly I pulled back the small vinyl curtain to expose the little window behind. We could see Malcolm lying on a stretcher covered with white sheets, only his face exposed. I burst into tears. So did my children. We sank into a corner, hugging each other and burying our heads against each other. I glanced over at Dolores. Her grief, as she beheld the body of her beloved son, seemed too great for either tears or whimpers. Inside, I knew that like all of us, Delores was experiencing hell. She stood rigidly, staring at the body of her son, with unbearable pain in her eyes. Then abruptly she turned and left the room.

As I looked back at Malcolm, I remembered the vivid dream that I had just two months before, the dream about Malcolm being trapped in an all-white attic by nameless colleagues and my frantic attempts to reach him. *Had that dream actually been a prophetic vision of what was to befall him? Did that vision offer clues about the cause of this sudden, unexplained death?* There was an unusual sense of peace after the storm about Malcolm now, after the ranting over the phone and running from what he feared. That fear was gone now, and he lay waiting for us to take him home.

I looked closer at Malcolm lying on that cold, hard surface. There were no visible signs of any spots or marks on his face offering clues to the cryptic nature of his death. I noticed the distended stomach, filled with liquid from his deadly infection. And I was struck by the large mouthpiece protruding from his mouth like some alien thing. It must have torn soft tissue as it was inserted down his throat, causing even more pain during his final agonizing fight for life. The pain slowly crept into me, and I started to sob uncontrollably at the sight. I tried to comfort myself with the thought that he was free of all pains now and with God, but the many unanswered questions surrounding his death sent my head spinning. *Had it been his life he was afraid for when he was yelling into the phone that day in July?* I didn't know, but I resolved, in that moment, to find out if I could.

"We May Never Know the True Cause of His Death"

Malcolm had a fiancée, Casia, who had lived with him. I could not help but wonder if she had experienced the same kind of emotional roller coaster, the hysterics and turmoil that had terrified Malcolm. Luckily, Casia was home when we returned from the morgue. Casia noted that Malcolm's body was all blue when she saw him moments after his death in the hospital. I was perplexed by his lack of oxygen and his body turning deep blue relative to the actual cause of death. There must be more to the story.

Minutes after Malcolm's passing, Casia had called Schneewind to help her cope with the death. Schneewind promptly came to the hospital. At Malcolm's deathbed, Schneewind examined Malcolm's body and told Casia, "We may never know the cause of his death."

His remark seemed a strange at a time when Malcolm had just died and before any autopsy was considered. Did he really think a diagnosis was so hopeless, or was he trying, for some reason, to dissuade the family from looking for the cause? If the latter, he had mistaken the effect of his statement, at least on me.

His words recalled a fable that I had heard as a child. When a rich local merchant wanted to take a vacation to a distant land, he hid his three hundred silver pieces in a wall near his bedroom. Afraid that the thieves would come and steal his fortune, he left a note by the wall saying, "There are no three hundred silver pieces in this wall." With this clever bit of misdirection in place, he left on his journey, confident that all would be well. Three months later, the merchant returned and rushed to the place where he had hidden his silver pieces. Ah! The three hundred silver pieces were gone. In their place lay a note that read, "Thank you for the tip."

Photo taken outside of a silversmith's store with a sign that says there are no 300 pieces of silver here

Was Schneewind's statement about not finding the cause of death, meant to turn away anyone intent on finding what lay beneath the surface? Was there something going on in the university's laboratory that the chairman didn't want questioned? It was a chilling clue that he had given us, but what had happened? Did he know something that we did not?

Great Lakes Regional Center of Excellence for Biodefense and Emerging Infectious Diseases Research—Yersinia pestis, a Biological Weapon

I opened up my laptop and found the following passages from an online article regarding the biodefense and bioterrorism project going on at the University of Chicago (http://www.glrce.org/research/vaccine.shtm).

Research Project 5—Vaccines against Plague

"*Yersinia pestis*, the highly virulent agent of plague, is a biological weapon. Strategies to prevent plague have been sought for centuries, however neither an FDA approved vaccine or the molecular basis of plague immunity are established. Immunization of animals or humans with live-attenuated (non-pigmented) *Y. pestis* strains raises protective immunity; however associated side effects prohibit the use of whole cell vaccines in humans." Figure 5A

Research Project 5 - Vaccines against Plague

Yersinia pestis, the highly virulent agent of plague, is a biological weapon. Strategies to prevent plague have been sought for centuries, however neither an FDA approved vaccine or the molecular basis of plague immunity are established. Immunization of animals or humans with live-attenuated (non-pigmented) *Y. pestis* strains raises protective immunity; however associated side effects prohibit the use of whole cell vaccines in humans.

Figure 5A: Mission Statement of GLRCE National Lab's Research Project 5–Vaccines against Plague

This passage gave me cold shivers. Schneewind's group had been working with an agent that caused plague and developing a vaccine that would protect against a plague infection. Had something gone wrong in the laboratory that Malcolm had become a victim of? I had no reason to think one way or another.

It is hard to describe how pain and grief can grip one's soul and make every normal function, like walking down the hallway, feel hellish. I could not even imagine what it would take to discover the truth behind Malcolm's death. But I had no choice; I owed it to Malcolm to find answers.

We returned to the Bernard Mitchell Hospital, where Brooke signed the request forms for an autopsy on her father. Malcolm's parents were also there.

Schneewind and his wife, Dominique Missiakas, came to greet us. This was the first time that I had met with the Schneewinds, who had come to the University in 2000, four years after I had left. I assumed that Schneewind knew of me from the years of legal battles in *Chou v The University of Chicago*.

After introducing himself, Schneewind asked me courteously to furnish a list of speakers for Malcolm's memorial, which he was setting up. I jotted down a few names and then glanced up. Schneewind, who had been watching me, immediately looked away. As the encounter continued, I realized that he was avoiding direct eye contact with me. He also avoided responding to any of my initial questions about Malcolm's death, turning away to greet my children and Malcolm's parents instead.

Malcolm's Last Email to Schneewind on 9/10/2009

That evening, I sat in the study where Malcolm used to sit. I saw his computer, printer, and stereo were left on and there were papers lying on his desk giving the impression that he had only temporarily stepped away and would return to them sometime later. "Poor soul!" I exclaimed. He had no idea that was his last day of life on September 13, 2009, when he went to the emergency room.

September 2009

		1	2	3	4	5
6	7	8	9	10	11	12
13	14	15	16	17	18	19
20	21	22	23	24	25	26
27	28	29	30			

I blinked away tears as I sat down before his computer. A message in his mailbox popped right before my eyes. It was Malcolm's last email sent to Schneewind and Bill Blaylock on September 10, 2009, three days before he died (Figure 5B)

Date: Thu, 10 Sep 2009 11:48:01 -0500
From: Malcolm Casadaban <mcas@uchicago.edu> Add To Address Book
Subject: Bad Flu
To: Bill Blayloc <BBlayloc@bsd.uchicago.edu>, Olaf Schneewind <oschnee@bsd.uchicago.edu>

```
I have come down with a very bad case of flu like symptoms:
Very high fever, muscle aches, and head aches with cloudy
thinking.
So I haven't been able to get into the lab and it is great
difficulty that I am even writing this email!

Hopefully I will get over these symptoms.
```

Figure 5B: Malcolm's last email sent to Schneewind and Blaylock three days before he died

It was sad and most gut wrenching to see Malcolm's last email sent to his colleague and chairman, Schneewind, and to one of the students, Bill Blaylock, an obvious outcry for help. I hurriedly flipped through all his recent emails in the inbox folder to look for any returned emails from the two, but there was none. And none from anyone who would come to his rescue three days before he died. My mind went blank, not sure what to make of the situation. Perhaps I was too afraid to find out more. Before I had time to dwell more on his emails, other materials on his desk drew my attention. Near his computer lay a copy of the *Biosafety Manual for Working with Yersinia pestis strain KIM D27*, July 13, 2004, University of Chicago, Department of Microbiology (Figure 5C).

Biosafety Manual for Working with Yersinia pestis strain KIM D27—"Exitus Lethalis" July 13, 2004

I ran through its content like a hungry child who saw food at the next table. There I encountered the name of the organism, *Yersinia pestis*, KIM D27, for the first time. Though not fully recognizing its significance at the time, I made a mental note that KIM D27 might be the centerpiece involved in Malcolm's death.

Dear God, I thought, horrified by the clear implication in his email. Malcolm had described himself as having "flu like symptoms," the exact words cited in the *Biosafety Manual,* "fever, dyspnea, cough and other flu-like symptoms."

Other symptoms that Malcolm described, "very high fever, muscle aches and head aches with cloudy thinking." were also in clear alignment with those described in the *Biosafety Manual.* Could Malcolm have known the nature and source of his infection by the symptoms he related to the *Biosafety Manual*?

I decided to run an online search on *plague Infection symptoms* to get a fuller picture and determine if Malcolm's reported symptoms matched them. I pulled up the following online presentation of symptoms associated with a plague infection.

- Sudden onset of fever and chills
- Headache
- Fatigue or malaise
- Muscle aches
- Abdominal pain, diarrhea and vomiting
- Bleeding through teeth (Septicemia)

It was clear that Malcolm had all these symptoms in his email. Casia reinforced my conclusions through her own observation of the last days while Malcolm suffered.

Feeling distraught and sick at heart, I continued to read the rest of the paragraph in the *Biosafety Manual.*

> Approximately 1–2 days after initial symptoms, exitus lethalis due to the generalized shock, intravascular hemolysis and respiratory failure can occur. Following localized infections with Yersinia pestis (needle stick wound infection), a swelling of regional lymph nodes with discoloration and fever is observed (plague bubos). The initial syndrome can progress to generalized infection and

> exitus lethalis within 3–5 days. (Emphasis added, Figure 5C)

I was struck with the term *exitus lethalis* in the *Biosafety Manual* and looked up its meaning in the English medical dictionary. It meant "deceased/died/passed away."

Three days after sending an email reporting his flu-like symptoms to his colleagues, Malcolm had died.

Clearly Schneewind was the PI who had to be contacted in case of emergency according to the *Biosafety Manual* (Figure 5C, 5D). But at critical time when Malcolm fell dreadfully ill on September 10, Schneewind and Blaylock both ignored him. No one from the university or the department ever came to his rescue. I felt disturbed by the tide of events leading to Malcolm's death. How could this be? This was the esteemed University of Chicago with topnotch scholars and faculties in a NIH designated national lab. My suspicions had to be wrong, no more than silly thoughts.

I hid myself in a corner in that little study upstairs; a ghostly eerie feeling that filled the room. I began to jot down all pieces of puzzles I had gathered this day.

1. "*Yersinia pestis*, the highly virulent agent of plague, is a biological weapon." In the Mission Statement used by the national lab of GLRCE in Bioterrorism Project (Figure 5A).

2. Mission Statement of GLRCE supported the notion of an immunization program of *human trial* using a strain of live-attenuated vaccine *Y. pestis* strain (Figure 5A).

3. Mission Statement of GLRCE supported a notion that there were side effects from the human trial that prohibited the use of *whole cell vaccines* in human (Figure 5A).

4. Malcolm appeared to have suffered a *plague* infection by the symptoms he had described in his email to Schneewind and Blaylock (Figure 5B).

5. It was apparent that Malcolm knew the source and nature of his infection as he described himself having as *"flu-like"* symptoms. It was the exact words used in the *Biosafety Manual for Working with Yersinia pestis strain KIM D27*, a plague agent (Figure 5B, 5C).

6. Malcolm sent his last email to Schneewind, knowing that *Schneewind was the PI and the person in charge in case of exposure (Figure 5C, 5D).*

7. *KIM D27 was the live-attenuated plague vaccine* in the *Biosafety Manual* to be used in human trials as in the Mission Statement (Figure 5A, 5C).

8. Infection by KIMD27 could result in "*exitus lethalis,*" which means death (Figure 5C).

The implications were heart chilling. *Who in his right mind would want to be voluntarily infected with a deadly plague agent such as KIM D27 that could result in exitus lethalis? And be left to die without rescue at the deathbed? But how else could it have happened? Could it have been deliberate, through a vaccine?* I felt sick at the thought. But I was more determined to find out if the plague bacterium, KIM D27, was indeed the culprit that killed Malcolm.

I had found the first solid clues about what had happened to Malcolm, about his possible infection by a plague agent and resultant death. My science training and experience, and my love for Malcolm, proved to be invaluable in my quest for truth. In a more spiritual sense, I felt Malcolm was near me, part of me, leading me on my fervent quest to seek the truth of his death.

I stayed in Malcolm's small study on the second floor of his house where I felt his presence and heard his woeful cry. I prayed fervently, "Malcolm, help me, I need more clues, clues that will set you free." I felt a new surge of energy running through me when I lay down to rest that night. Malcolm was there in his little study, giving me the strength and courage I so needed to sustain my investigation.

Biosafety Manual for Working with *Yersinia pestis* strain KIM D27

The University of Chicago
Department of Microbiology
071304

II. Reporting *Yersinia pestis* infections

Despite the fact that *Yersinia pestis* strain KIM D27 is an attenuated strain, if the following symptoms arise the laboratory worker must report to Occupational Health Medicine (OHM) room L156 phone# 2-6757 or Bernard Mitchell Emergency Room, if OHM is not open, immediately. Symptoms of Yersinia pestis infection are as follows: A nonspecific syndrome (i.e., fever, dyspnea, cough, and flu like symptoms) follows inhalation of infectious Yersinia pestis bacilli. Approximately 1-2 days after initial symptoms, exitus lethalis due to generalized shock, intravascular hemolysis and respiratory failure can occur. Following localized infections with Yersinia pestis (needle stick wound infection), a swelling of regional lymph nodes with discoloration and fever is observed (plague bubos). The initial syndrome can progress to generalized infection and exitus lethalis within 3-5 days. More information about the symptoms and signs of plague infections can be obtained at the CDC web site for bioterrorism (http://www.bt.cdc.gov).

A licensed vaccine is not presently available in the U.S.

Dr. Schneewind, must be informed of any laboratory accidents and exposures. The PI is available on a 24 hour basis (via cell phone at 773.824.2628). Occupational Health Medicine during regular business hours will fulfill the needs of laboratory workers as the primary contact after accidents and exposures. The medical emergency room on East 58th street is responsible for treatment when Occupational Health Medicine is not open.

Figure 5C: Biosafety Manual For Working with Yersinia pestis strain KIM D27

I. Study Contact Information:

PI Name: Olaf Schneewind		Unique ID #: 10811
Appointment: Professor, Principal Investigator		Department:
Office Address: 920 E. 58th street		E-Mail: oschnee@bsd.uchicago.edu
Building: CLSC	Room #: 607	Mailcode: 60637

In case of an emergency, contact the PI at:		
Phone: 4-9060	Pager:	Home Phone: (b)(6)

Oversight/Alternate Contact: In the absence of the PI, who will assume responsibility for the ongoing day-to-day oversight and supervision and personnel.		
Name: Lauriane Quenee	e-Mail: lquenee@bsd.uchicago.edu	
Phone: 4-0566	Pager:	Home Phone: (b)(6)

Figure 5D: Schneewind, PI of the GLRCE was to be contacted in case of emergency

6

THE TURNING POINT

On September 16, 2009, we gathered to honor Malcolm at a memorial service, organized by Schneewind, at the University of Chicago's Bond Chapel. Although the mood was somber and grief stricken, music played by the organist of Rockefeller Chapel soothed our hearts. Beautiful stained glass windows enclosed us; sculpted angels, wings outspread, perched above us, playing instruments; and the words of the Beatitudes, carved into walls on either side of us, blessed us. Later, as events began to build, I would think of one in particular: "Blessed are they that do hunger and thirst after righteousness."

Family and friends came forward to the altar to speak about Malcolm. Brooke and Leigh, Malcolm's beloved daughters, spoke first. Through their tears, they shared childhood tales as well as more recent stories that clearly showed their enduring love and respect for their father and his unconditional love and support for them.

Casia then spoke of the kind, loving fiancé that Malcolm had been to her and her fond memories of him in the last two years of his life. Malcolm's brothers and sisters followed, their love for their oldest sibling and the pain at his loss clear to see on each of their faces.

Then Bernie Strauss stood up to give his eulogy. I had known Bernie since Malcolm and I had come to the university in 1981. Bernie was a senior faculty member and a departmental chair, and at one time, a dean at the university. Because of his research focus on bacterial genetics, he came to know Malcolm well as the two spoke a common language.

Bernie told several stories about Malcolm; one in particular, which occurred one day in 1985, stood out because it said so much about how the wider scientific community viewed Malcolm. That day in 1985, Bernie had been hosting a group of NIH grant reviewers who had come to the university for a site visit. During the visit, one NIH official asked whether Malcolm had received tenure yet. Academic tenure was usually awarded to young faculty members six or seven years after their entry level at assistant professor. This was the fifth year for Malcolm.

This NIH official then told an inspiring story of his interaction with Malcolm. He marveled at Malcolm's ingenuity in creating the gene fusion procedure, so powerful, so transforming for scientific research, and how grateful he had been that Malcolm had taken the time to help him solve his experimental difficulty—as if it had been Malcolm's own research.

The NIH official then said jokingly that if Malcolm had not gotten tenure from this university, perhaps the NIH would consider deferring the central grant award until Malcolm received his due tenure.

Bernie told Malcolm the story years ago, and Malcolm was pleased, though he could not recall who this person was because he regularly answered more than ten calls a day from scientists asking for his help in their own quests.

Shortly after that NIH site visit, Malcolm received tenure as an associate professor in the Department of Molecular Genetics and Cell Biology. Bernard Roizman was the temporary chair granting his tenure. It was the same year that I received my Ph.D. in virology for my work on herpes simplex virus in Roizman's lab. Jonathan Kans, who marched behind me in graduation procession, graduated from Malcolm's laboratory at the same time. Those were the good old days indeed. As I listened to Bernie talk, I wished I could literally step back in time, if only for a little while.

Following Bernie Strauss's eulogy, two of Malcolm's former students—Eduardo Groisman, who had gone on to become a faculty member at Yale University, and Dave Demirjian, who became the president of two biotech companies—rose to talk about their former professor and mentor. Malcolm loved teaching and helping young scientists along the way. In fact in 1990 he broadened his teaching efforts from the university to neighboring institutions, where he frequently became the student adviser and mentor for their Ph.D. theses.

Midwest Journal Club

There was a time when Malcolm, along with his peers at the University of Illinois at Chicago, formed a Midwest journal club to promote prokaryotic science. Scientists from Northwestern University, the University of Illinois at Chicago and Urbana-Champaign campuses, Loyola University, Purdue, Washington University in Missouri and others from Wisconsin and Michigan would gather on a Saturday afternoon for a day of seminars. The club would meet one afternoon every third month on the first floor of the Cummings Life Science Center, room 101, at the University of

Chicago. The club was to promote scientific exchanges among all local labs and to broaden and enhance the educational trainings of students who were interested in joining the various disciplines.

In early days, the club was set up and financed by various federal and state agencies and sometimes by local institutions. The seminar would include three one-hour presentations by reputable scientists preceded by a buffet lunch.

In the evening, most faculties and students who came to Chicago would go out to local Chicago bars and restaurants to enjoy a night out before driving back to their home states. It was such a wonderful educational program attended by many who would battle wintry storms to come to Chicago to enjoy a day of science in good company with Malcolm and other scientists.

Malcolm played an important role in organizing such activities that went on for several years. He thoroughly enjoyed the science, the company of great scientists and colleagues that he used to know. They would discuss science, a movie, or a book without worrying about the politics behind it. He would even bring our daughter Leigh along to the seminar where he bestowed upon her the honor of lighting at the seminar. As Leigh flipped the light switch in Cummings room 101, the seminar began. Malcolm loved the scientific exchanges of this kind; it was one of the highlights of his life.

In the 1990s support funds for the program dried up. To keep the club going, Malcolm picked up the expenses himself. In those days, in order to save money, Malcolm volunteered me to do the lunch buffet. So we would order Kentucky Fried Chicken as the main dish and I would supply sides from our home kitchen, such as salad, Cajun rice, and potato salad. Our children would get drinks and coffee ready for the crowd. Though the menu stayed simple and economical, a crowd of over seventy local professors and students would pack regularly into Cummings 101. Perhaps people liked to visit Malcolm or simply liked to visit Chicago for a day and a night out, but for whatever reason, it was a successful event time after time.

Gradually, despite all Malcolm's efforts, the event failed as financial hardship continued to take its toll on the scientific community. I also had less time to help as I became fully engaged with my own research. The Midwest Journal Club finally ended in 2000.

The final tribute to Malcolm came from Schneewind himself, who delivered a solemn eulogy at the end. He spoke kindly and with reverence about the man he had worked with for so many years. As if they had been the best of friends. And yet this supposed friend had not responded to Malcolm's email about his flu-like symptoms and his cloudy thinking, had not met my eyes or answered my questions at the hospital.

The Reception—Undercurrents

The memorial reception was packed with people who came to say good-bye to Malcolm. I knew many from before I left the University of Chicago in 1996.

Ling Chan was the first to greet me as we streamed out of the chapel. She spoke to me in Chinese, emphatically saying that the people in the Department of Molecular Genetics and Cell Biology were very mean to Malcolm, that they had been after him, retaliation for my lawsuit. She then claimed that Schneewind was the hero who had rescued Malcolm from the harmful environment, had played the part of savior in Malcolm's life.

I didn't trust her words. Ling had tried to cause a rift between Malcolm and me just months before. She must have thought I didn't know, that Malcolm wouldn't have told me that she visited him and accused me of having an affair some twenty years before. She even hinted that Malcolm might not be the father of my children. Here was a woman who was accusing others of persecuting Malcolm when she herself had spoken lies to him at a time when he was beset by troubles within his department and worried about the university fraud committee's trumped up investigation.

I turned away from her and moved to greet others. I came face to face with a faculty member who had been a friend of Malcolm's for a long time and spent time with both of us. We knew each other well. But instead of offering condolences, he threw his hands up, as if in a gesture of surrender, stepped back, and shook his head, as if saying, "I did not do it. I had nothing to do with this. And I have nothing more to say."

It was a startling act from a man who worked with Malcolm and knew us both, who should have been offering condolences. I moved away from him since our exchanges were awkward at this point. Unfortunately, his would not be the only strange reaction.

I also found it strange that Malcolm's former student Claire Cornelius did not come to spend time with the family. I had heard about her from Casia. That had been in response to my question to Casia about why Malcolm had been wearing his Harvard regalia in the handsome photo of him that appeared on memorial announcement (see cover photo). I noted a hint of sadness in his eyes, and I asked Casia when and why he wore such formal academic attire in academic gown and hood?

Casia said he had worn the regalia to march in a graduation procession to honor his student Claire Cornelius. She had just gotten her Ph.D. in June of 2009, three months before Malcolm died. Claire had planned to graduate in June quarter, but for unforeseeable reasons, had moved her graduation ceremony to a later day but forgotten to tell Malcolm. Casia apparently had met Claire on several occasions and knew her well. The following was taken from:

http://www8.nationalacademies.org/cp/committeeview.aspx?key=48923

Major Claire Cornelius
U.S. Department of the Army

Major Claire Cornelius, D.V.M., is currently pursuing a Ph.D. in microbiology as part of a U.S. Army Long-Term Health Education and Training (LTHET) opportunity at the University of Chicago. Her doctoral thesis research interests include plague pathogenesis and vaccine design. Before beginning her doctoral studies, she served as Post Veterinary Officer, Guantanamo Bay, Cuba; Force Veterinarian, Multinational Forces and Observers, Sinai Peninsula with duty in Egypt and Israel; and Branch Chief, Yokosuka Branch Veterinary Services, Japan with additional duty in Thailand, Philippines, Vietnam, Hong Kong, and Indonesia in support of public health and civic action programs. She has completed the Foreign Animal Disease Diagnostician Course (FADD) at the Animal Disease Center/USDA located on Plum Island, New York and the Summer Institute in Tropical Medicine and Public Health at the School of Hygiene and Public Health at Johns Hopkins University. Additionally, she has been a member of two research teams investigating malaria and/or hemorrhagic viruses in the Amazon basin-Iquitos, Peru

Figure 6A: Career experience of Army Major, Claire Cornelius, in Chicago National Lab

What I found out about Claire represented a different image than I had anticipated. She was a U.S. Army Major from the DOD (Figure 6A, emphasis added). I had never thought it possible that a student in the laboratory for whom at graduation Malcolm wore his Harvard regalia would be a military officer employed by the DOD. It surely raised an alarm in my mind. I knew the DOD had conducted human trials in the past. Why had this officer from the DOD been in Malcolm's lab?

After the reception, the crowd began to move back to their own buildings to continue work that afternoon. As people dispersed, I saw another faculty member who had been a colleague of both of ours, whom Malcolm and I had known since our first day in Chicago. I went forward to greet him, glad to see him after such a long time, but he quickly turned away and headed off in the other direction. I tried to catch up with him, but lost sight of him in the crowd of moving people.

When I finally saw him again by the elevator, I gathered Malcolm's parents and his sisters to meet him. "Here is someone you should meet," I said calmly to the Casadaban clan. "He was the one who helped us a lot when we first arrived at the university."

"Yes, yes," said this colleague and friend of Malcolm's. "Malcolm was a good man. A brilliant man."

He then said abruptly, "Excuse me, but I have to go. It was nice meeting everyone." Just like that, this friend of many years disappeared into the crowd. He never once looked me in the eye.

There was something very wrong here. If Malcolm had died of natural causes, if there was no mystery about his death, people would have greeted the family with unreserved compassion, offering words of condolence and hugs of sympathy. Yet many of them acted as if the last place they wanted to be was here, talking to Malcolm's family, talking to me.

I sank in my seat. Desperation came upon me: It was like a masquerade party that I had dreamed about on the plane to Chicago. Everyone wore masks. There were people laughing, sneering, and joking; there were trumpets playing and angels singing, but I could not find Malcolm. Such was the crowd who gathered to pay tribute to a man, their colleague and friend, and a professor at the university for more than thirty years. Sad as I was, I had no time to spend worrying about their unnatural behavior. My priority was to find the truth.

The Turning Point—Unidentified Bacteria

Brooke came to inform me that the family had just arranged to meet with the doctor who had treated Malcolm in the emergency room in the final hours of his life. Brooke rushed us to the Cummings Conference Room on the eleventh floor. I sincerely hoped that Malcolm's doctor would explain what happened to Malcolm to us, the family, since only the family members would be truly grieving over the loss of a loved one. There would be no one there with a vested interest in shading the truth.

In the conference room, I sat around a long table with twenty members of the Casadaban family members, all anxious but sad and afraid to hear the grim details about the final hours of Malcolm's life. I intentionally took a seat next to the doctor so that my questions or comments could clearly be delivered and heard by him.

Dr. Howes began by telling us that he had known Malcolm personally from the many years they had both been at the university, and in fact he had requested this meeting with the family alone as a tribute to Malcolm. Dr. Howes slowly recounted the tragic moments of Malcolm's final hours. We felt deeply grateful to this man who had taken good care of Malcolm when none of us could be there.

"Malcolm had so much bacteria in his blood that his blood serum looked turbid to the naked eye in the emergency room," Dr. Howes explained. "It had become apparent that Malcolm was fighting a losing battle. Despite this, we wanted to do everything we could to save him, so we gave Malcolm a strong dose of antibiotics at around 11:00 a.m. This was eight hours after he arrived in the emergency room."

Dr. Howes noted that human blood typically was like a sanctity pool and holy water. A trace amount of bacteria, just a few molecules of pathogen in human blood, would set off a full-blown blood infection leading to death. That, Dr. Howes said, was what happened to Malcolm. He suffered a full-blown infection in his blood called septicemia; the death was inevitable, a result of the bacterial toxins released to the tissues.

Dr. Howes went on to explain that Malcolm's blood contained two other bacteria species that had already been identified to be of streptococcus origin and *Klebsiella pneumoniae*. As medical staff analyzed Malcolm's serum, they knew there was no chance that he could survive the infection. Dr. Howes folded his hands on the conference table and looked down. We could see the deep sorrow that he had not been able to save Malcolm.

Hearing about Malcolm's final struggles was gut wrenching; everyone was crying, some muffled sobs. I wiped my face, pushing my sorrow aside to concentrate on whatever else Dr. Howes might tell us.

"However," Dr. Howes added, "There was also an abundance of a gram-negative bacteria, which the doctors failed to identify at that time."

Gram-negative bacteria? I wondered.

Dr. Howes paused and said, "On an educated guess, do you, any of you, know what that gram-negative bacteria is?"

I leaned over and whispered, "*Yersinia pestis.*" I knew that was the bacteria Malcolm had been working on in his lab. I also knew from the *Biosafety Manual* that was in Malcolm's little study marked the pathogenesis of KIM D27 with *exitus lethalis*. In addition, Malcolm had all the symptoms of a plague infection.

Dr. Howes shook his head immediately. He clearly did not want to entertain my idea. Perhaps he did not know that I was also a scientist. Perhaps Dr. Howes had consulted Schneewind already on that night when Malcolm died and was told that the cause of death might never be found.

He continued to speak for a few more moments when I decided to throw out the idea one more time—this time a little louder. "*Yersinia pestis*," I said loudly, so the whole room would hear it this time.

Dr. Howes frowned and shook his head as though I were a lunatic. The way he brushed my observation off, he must have thought I was just making some wild guesses based on something I had read or heard. I could tell by his facial expressions that he wanted me to shut up. He must have thought I was crazy to even imply such a monstrous notion as the plague! After all, Malcolm did not seem to have any external appearance of a plague-inflicted death.

I realized that Dr. Howes probably had never seen the *Biosafety Manual* and other information that I found in Malcolm's study and did not know what Malcolm had been working on. Yet his persistent refusal to even entertainment what I was telling him without asking where I was getting my information began to get under my skin. After all, weren't we all trying to discover the truth about what killed Malcolm?

Dr. Howes finally brought up concerns about Malcolm's dental cleaning since he had noticed some bleeding of Malcolm's gums in the ER. So now everyone, including me, wondered if some kind of gram-negative bacteria might have gotten into Malcolm's blood through dental cleaning. But I also knew, having researched that symptom, plague-inflicted septicemia patients usually would exhibit some gum bleeding in late stage of the infection. The mystery gram-negative bacteria had to be a blood-borne pathogen since it had grown to confluence in Malcolm's blood.

Given the information I had uncovered, I felt I had to insist that *Yersinia pestis* could be the unidentified bacteria and it should be explored as a possibility in the autopsy. It should at least be ruled out scientifically. So once more I said, "*Yersinia pestis*."

Dr. Howes paused and looked over at me coldly. But then I saw a change in his face, as if he was finally considering the possibility, perhaps considering Malcolm's symptoms in light of a possible plague infection.

So I said a fourth time, "*Yersinia pestis*," adding, "Malcolm worked with *Yersinia pestis* in the lab."

The doctor still looked uncertain, and I knew we would have to push for more if we wanted to find the truth. I told Dr. Howes firmly, "We're going to have our own autopsy done to test for *Yersinia pestis*."

Dr. Howes was caught off-guard by my comment, but rather than respond, he changed the direction of the conversation.

On that September 16, 2009 at 7:30 p.m., Dr. Howes ended our family meeting after an hour and half of stimulating debates. The meeting left the family with more questions than answers. Had Malcolm been exposed to some kind of bacteria from dental cleanup? What about the gram-negative bacteria that Dr. Howes had found that grew in Malcolm's blood? What strain was it? Did he have an open wound where the organism might have entered? Was there an accidental spill nearby? Did he inhale the deadly bacteria? What about the vaccination programs that people in that lab had to undergo before entering?

In truth, I had very little evidence to go on at this point. What I knew was that Malcolm had worked on a strain of *Yersinia pestis* and had symptoms associated with plague infection. But there was the strange behavior of the university people before and during the reception to consider. Strangely no one from Schneewind's laboratory ever approached the family to offer condolences at the reception. The behavior exhibited by Malcolm's friends and colleagues at the memorial and reception did not make sense unless they knew, at least some of them, that there was something very wrong about Malcolm's death.

Let the Truth Be Told . . .

Over the next two days, the family was so preoccupied with paperwork, medical records, and the funeral service arrangements that there was no time to think about anything else. After the funeral, Malcolm's siblings began to leave town to return to their homes.

On Friday, September 18, 2009, around 5:40 p.m., my children and I were scheduled to pick up Malcolm's belongings from Schneewind's lab on the sixth floor of Cummings Life Science Building. It had been arranged several days beforehand. Exhausted and sick at heart, I felt reluctant to go up to the lab to face students and researchers who had worked with Malcolm in the past few years. I decided to wait for my daughters in the car while they went into the lab.

Within minutes, I received a frantic call on my cell phone from Brooke, who was sobbing. It was as if she had been hit with another dramatic event that had sent her into hysterics. I listened carefully, but the words were readily not comprehensible.

"Brooke, listen to me. Was it *Yersinia pestis* in Malcolm's blood?" I asked, the first question that leapt to my mind. I remembered well how I pushed Yersinia pestis to be tested by the autopsy lab in our meeting with Dr. Howes.

"Yes, yes!" she shouted. "Mom, come up here quickly. We're in Dad's lab and the students told us that Dad died of *Yersinia pestis*! Hurry! ... Hurry! ..."

I could hardly believe what I was hearing. Suspecting was not the same as knowing. But finally, the truth was coming out.

7

"MALCOLM HAD JUST BEEN VACCINATED"

On September 18, 2009, at 5:30 p.m., I rushed out of my car and ran into the Cummings Life Science Building to join my daughters on the sixth floor. Entering Schneewind's lab, I found a crowd of students surrounding Brooke and Leigh, both of whom were crying. Some students were overcome with grief and started sobbing. I could see in their expressions and by their actions that they cared deeply for Malcolm.

I heard the term *KIM D27* whispered a number of times. I remembered it from the *Biosafety Manual for Working with Yersinia pestis strain KIM D27* in Malcolm's house. I sensed that my persistent discussion with Dr. Howes just two days ago had finally brought out some answers. I could not wait to hear some confirmation words from these students in Schneewind's laboratory.

Bill and Juliane

Soon after I entered, my children and I were ushered to a private room with Bill Blaylock and Juliane Wardenburg. Juliane, who held an M.D. and a Ph.D., had once worked in Schneewind's lab and then become a faculty member in the Department of Microbiology. Bill was a senior scientist, a postdoctoral researcher in Schneewind's lab. I knew that Bill, who occupied a bench next to Malcolm's in the lab, had worked with Malcolm on various scientific experiments.

Now the plague bacteria had surfaced, the question that raced through my mind was what strain was it? Could it be the KIM D27 that I had read about in the *Biosafety Manual*? How was it delivered to Malcolm? All that we had been told so far was that the bacteria had been identified and all Malcolm's personal belongings from the lab were now under quarantine. Nothing could be removed from his office or the lab bench.

I also knew that Bill, like Schneewind, had been aware of Malcolm's symptoms at least three days before his death and had offered him no help. It seemed unprofessional, perhaps even a direct violation of basic principles of biosafety, principles set up and overseen by the IBC. Was a gross biosafety policy oversight rocking the city of Chicago, the university, the Cummings Life Science Building, and the many researchers and staff engaged in university bioterrorism research?

Bill told us about the organism that had invaded Malcolm's body and killed him, though he did not know whether it was the wild type plague strain or the one that had been attenuated for the study. In fact at that moment, Schneewind was doing the PCR (polymerase chain reaction) experiments next door to make that determination.

According to Bill and Juliane, Schneewind had spent the whole afternoon meeting with university administrators, legal counsels, deans, department chairs, and infectious disease experts to deal with the new crisis of a university plague-inflicted death. I could imagine the tense, perhaps near-panic mood of these meetings as the participants wondered: Was the university's reputation at stake? Would the grant money dry up? Would the laboratory be shut down?

Bill educated us on the three variable forms of plague: bubonic, pneumonic, and septicemic. Malcolm apparently suffered a septicemic form of plague characterized by the infection of the blood consistent with what Dr. Howes had described just two days ago.

I asked, "Bill, what would be the symptoms associated with a plague infection?"

Bill replied without hesitation, "The symptoms are flu-like."

It was the same description of symptoms found in the *Biosafety Manual* (Figure 5C) and in Malcolm's email (Figure 5B) on September 10. Because of my background in microbiology and virology, the science and medical language made perfect sense to me. My next question was, what plague strain was it?

Live Attenuated Vaccine Strain, KIM D27

Bill went on to describe the genomic structure of the KIM D27 as a genetic variant of the wild type strain (KIM strain). It was the strain composed of a 102Kb genomic deletion in a region of the chromosome called high-pathogenicity island. This region was clustered with elements that cause plague pathogenesis such as one involved in bacterial iron uptake and the pigmentation segment (pgm^-) associated with Congo red staining. Because of the deletion in the pathogenicity island, the KIM D27 strain retained a weakened ability to cause illness in animals. Its replicating potential, however, could elicit an immune response suitable for the vaccine delivery.

The dual properties associated with KIM D27 made it the live attenuated plague vaccine that we discussed in this book.

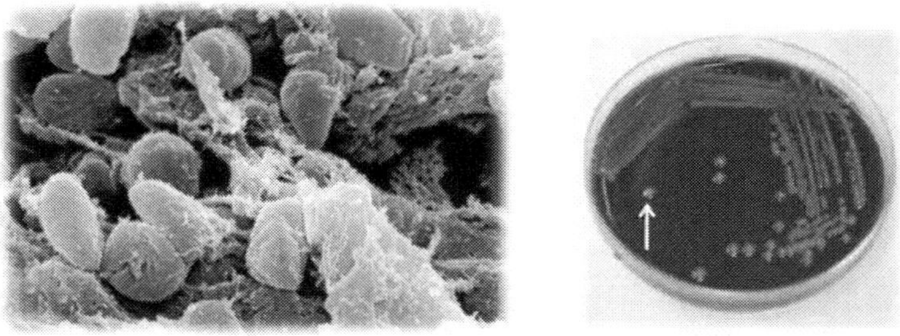

Figure 7A: (Left): Electron micrograph of KIM D27 bacteria. (Right): $\sim 10^7$ bacteria in a colony shown with arrow.

KIM D27 Vaccine and Virulence

Bill further enlightened us on the property of KIM D27—its potential use as vaccine. He summarized the properties diligently as if from a textbook:

1. The organism is alive and can multiply by itself in the host. Naturally, while multiplying, it has the ability to elicit host immune responses, as a vaccine does.

2. The strain is attenuated with a reduced virulence—that is, the ability of the strain to cause disease is weakened—as compared with the virulence of the wild type. However, as neatly tucked away as it seemed, the strain's reduced virulence in subcutaneous (beneath the skin) infection is not the same in septicemic (blood) infection. In the blood where there is a steady state of iron influx to enable bacteria multiplication, a few attenuated bacteria that enter successfully could multiple to a devastating level and end in *exitus lethalis*, a 100 percent fatality rate.

3. Thus the mechanism of delivery, portal of entry, to the host ultimately determines the virulence potential and the fatality rate. Unknown to us at the time, Schneewind's research efforts in 2006–2009 were largely dedicated to optimizing a route of delivery to

bring KIM D27 to the blood. After all, that kind of target delivery is the very essence of Biowarfare.

Bill further explained the virulence potential of the plague strain. The deletion in the genome of KIM D27, the live attenuated vaccine strain, rendered a degree of 10^7 fold of reduced virulence. In layman's terms, medical designation of $LD_{50}=10^6\text{-}10^8$ in mice suggested that 10^7 colony forming units (Figure 7A) of KIM D27 when inoculated to mice would present clinical symptoms (ruffled fur and lethargy) at 30 percent mortality rate tested (Figure 7B). Wild type plague, in contrast, would require 2-4 bacteria to kill 50 percent of the mice tested ($LD_{50}=2\text{-}5$), a 10^7 fold enhanced pathogenicity over the live attenuated vaccine strain. LD_{50} was a nomenclature for lethal dosage (LD) defined as the amount of bacteria that would kill 50 percent of the animals. I believe Bill was referring to the article by Quenee et al. (2008) in *Infection and Immunity*, *"Yersinia pestis caf1* Variants and the Limits of Plague Vaccine Protection" (http://www.ncbi.nlm.nih.gov/pmc/articles/PMC2346721/)

RESULTS

Plague immunity generated with live, attenuated *Y. pestis* (Δ*pgm*) strains. We examined the virulence and vaccine attributes of *Y. pestis* KIM D27 (biovar Medievalis), a Δ(*pgm*) strain harboring all three virulence plasmids (pCD1, pMT1, and pPCP1) (18) (Fig. 1A). Groups of 6- to 8-week-old BALB/c mice ($n = 10$) were immunized by intramuscular injection with 1×10^5 CFU and 1×10^7 CFU of *Y. pestis* KIM D27 suspended in PBS or with PBS alone. Animals were monitored for morbidity and mortality over the course of 21 days. In contrast to animals injected with PBS (all of which remained healthy), mice immunized with 1×10^7 CFU of *Y. pestis* KIM D27 presented clinical symptoms (ruffled fur and lethargy) and 30% mortality (Fig. 1B). Animals immunized with the lower dose (1×10^5 CFU) presented similar symptoms; however, all mice recovered 6 to 8 days postimmunization (Fig. 1B). To avoid vaccine mortality, the sublethal immunization dose (1×10^5 CFU) was chosen for future experiments.

Figure 7B: Reduced virulence of KIM D27 in Queene et al. 2008

Targeted Delivery of Pathogen

Having explained the amount of bacteria that would cause death in animals, Bill went on to lend us more insight on the delivery mechanism in which bacteria entered the host via its portal of entry. Ultimately, it was how and where the bacteria entered the host that dictated the outcomes of diseases. I was fascinated with his words and teaching. I had to fill in some gaps quickly to understand the new plague biology in order to get the whole truth on how Malcolm had died.

Bill said, "*Yersinia pestis* is a natural blood-borne pathogen." If the pathogen was delivered subcutaneously (below the skin) to a peripheral tissue site, the bacteria would multiply inefficiently. The inability to cause an infection could be, among all reasons, due to insufficient iron to support its infection at the peripheral sites and the rate of successful infection leading to the disease would be low, in the range of 50 percent fatality. In the context of intranasal delivery through breathing, the bacteria would have to multiply in the lung and enter the blood as in pneumonic plague. I remembered Dr. Howes just two days before told the family that Malcolm had a full-blown blood infection—septicemia—from the amount of deadly bacteria that were present in his blood.

Throughout this discussion, I could not help but wonder if the plague bacteria had been introduced into Malcolm's body intravenously, that is, through a needle to his vein. But how would that have come about? Accidentally during his research? *Highly unlikely,* I thought. Malcolm was a smart and careful scientist when he worked, taking all precautions to protect himself from the deadly organisms he handled. He would know how to protect himself from such intrusion. Besides, the doctors in the ER would have taken notice of any needle sticks or scabs formed around entry sites when they examined him for cause of death.

In my mind, I raced through the portal of entry mechanisms: *intramuscular* (into a muscle), *intranasal* (through the nose), *intravenous* (through the veins), *intraocular* (though the eye), *intragastric* (through the stomach and intestine), *oral* (by mouth). *How else?*

I believed Bill was talking about the portal delivery protocols with respect to the disease outcomes in the animal study in 2006–2008 in Schneewind's laboratory (Figure 7C, Appendix 2: Plague at the University of Chicago).

- *Intravenous* delivery to mimic *septicemic* plague infection
- *Subcutaneous* delivery to mimic *bubonic* plague infection
- *Intranasal* delivery to mimic *pneumonic* plague infection

Mouse model of plague

- Mice will be infected intravenously to mimic septicemic plague (dose: from 10 to 10^8 cfu, volume 100ul; location: retro-orbital sinus; manipulation: mice are anesthetized for the procedure, bacteria are injected with 28g needle).
- Mice will be infected subcutaneously to mimic bubonic plague (dose: from 10 to 10^8 cfu, volume 100ul; location: inguinal fold; manipulation: mice are hand restrained, bacteria are injected with 28g needle).
- Mice will be infected intranasally to mimic pneumonic plague (dose: from 10 to 10^8 cfu, volume 20ul; location: nostril; manipulation: mice are anesthetized for the procedure, bacteria are pipetted into nostril).

Figure 7C: Animal study of pathogen target delivered to mice that yielded various disease outcomes

Indeed, the portal of entry, or the site of pathogen delivery, marked the disease outcome (Figure 7C, Appendix 2). That is to say, the way in which the plague bacteria entered the body determined what kind of plague infection the person fell victim to. I came away with the understanding that Malcolm suffered a septicemic infection. That meant that there had to be a mechanism that allowed the pathogen to be delivered to his blood. But the doctors had found no needle marks or scabs that could indicate an injection. If not by direct injection into the blood, how else could the pathogens have been delivered? I could not easily decipher a likely route in reasonable terms to explain the cause of death in Malcolm's case.

Much of these had borne images out of my own experience in a herpes virus study in 1990 in an article published in *Science* on November 30, 1990, volume 250, page 1262 (Figure 7D). It was of the same kind of study done in the brain of mice to determine neurovirulence capability of the virus. This article addressed the specificity and the virulence potential associated with herpes simplex virus mutants, $\gamma_1 34.5$, live attenuated vaccine virus strain that I had constructed earlier. The neurovirulence was tested in the brain of mice by intracranial inoculation.

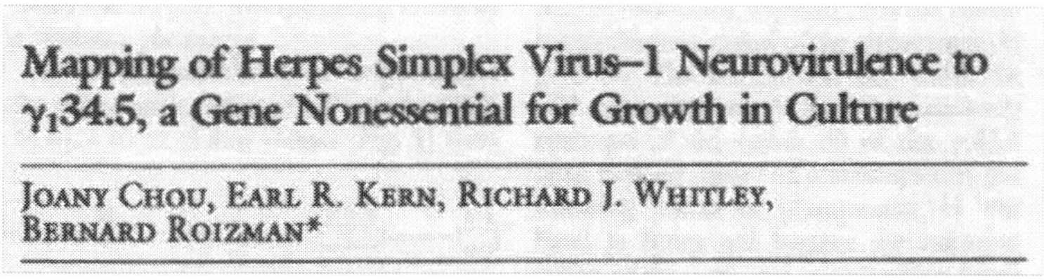
Mapping of Herpes Simplex Virus–1 Neurovirulence to $\gamma_1 34.5$, a Gene Nonessential for Growth in Culture

JOANY CHOU, EARL R. KERN, RICHARD J. WHITLEY, BERNARD ROIZMAN*

Figure 7D: Neurovirulence Property of $\gamma_1 34.5$ Protein in Herpes Simplex Virus I

Bill spent a good half hour helping us to understand this cryptic organism. His talk about the plague was most informative and illuminating.

Juliane Wardenburg had remained silent throughout our discussion. When Schneewind finally became available, we were led to his office, accompanied by Wardenburg. But before we left the student conference room, Bill told us one more fact, another shock in itself.

"Malcolm Had Just Been Vaccinated"

We had gotten up to follow Wardenburg into Schneewind's office, when Bill said, "Malcolm had just been vaccinated."

I turned back and stared at him blankly for a moment, as it seemed a non sequitur. Then I thought about the vaccines of H_1N_1 that came to mind first, since the so-called swine flu and the vaccine for it was in the news. But why would Bill mention an H_1N_1 vaccination now?

"What vaccine?" I asked.

Bill shrugged a shoulder as if to say 'Isn't it obvious?' But seeing that I still did not understood, he added, "It is a lab policy to have new people coming to the lab to be vaccinated."

Shocked, I realized he meant a plague vaccine. I would later discover that an IBC policy approved by the university administration and legal counsels on September 30, 2004, and still in effect required vaccination if that was the policy of the lab. Schneewind required plague vaccination for researchers in his lab. That said, researchers in his national lab are human guinea pigs for the vaccine testing. If not vaccine, biological warfare agent or any agent under disguise of vaccine was used.

Mandatory Vaccination at GLRCE National Lab

Even though there was a policy in place, Malcolm should not have been affected by this policy, I kept reassuring myself on this point. Realistically, Malcolm was a university-appointed faculty and an IBC biosafety officer who was intimately familiar with the biosafety standards of pathogens in the laboratory. He certainly was not an entering student in status; nor was he likely to have voluntarily subjected himself to the mandatory vaccination applicable to Schneewind's laboratory employee and staff only.

How then did Schneewind and members of the Microbiology Department overcome this hurdle to bring Malcolm to the human trial that he knew was a vaccine?

Suddenly a university IBC document that was in Malcolm's room gave me the clue I was seeking. The notion of a "new member" to Schneewind's laboratory also signified one's obligation to participate in the mandatory human vaccine trial that had been in practice in Schneewind's laboratory since 2004 (Figure 7E).

> "The University has concerns about mandatory vaccination. However, if the philosophy of the lab is to require vaccination, employees that wish to work with the agent would have to be vaccinated. If an employee refused vaccination, the University's legal office should be consulted, and alternative responsibilities would need to be assigned to the employee. " (Figure 7E, Appendix 6)

The Committee discussed vaccination of staff members working with *Bacillus anthracis*. The University has concerns about mandatory vaccination. However, if the philosophy of the lab is to require vaccination, employees that wish to work with the agent would have to be vaccinated. If an employee refused vaccination, the University's legal office should be consulted, and alternative responsibilities would need to be assigned to the employee.

Figure 7E: Mandatory vaccination at the University of Chicago, excerpt from University IBC

Lo and behold, in a Protocol Submission to NIH on October 8, 2008, Malcolm had been certified to become a new member to join Schneewind's bioterrorism laboratory. And the protocol was certified and approved by Lenora Hallahan, administrator for Regulatory Compliance on August 21, 2008 (Figure 7F, Appendix 4)

The specious attacks on Malcolm made no sense unless a noxious false agenda had been at work. In the same Protocol Submission to NIH on October 14, 2008, Malcolm had been downgraded from his academic tenure faculty to that of a research associate, an employee in status, in Schneewind's laboratory (Figure 7G, Appendix 2).

This downgrade compromised Malcolm's standing at the

university and made him a guinea pig in the human trials that doomed his life.

The Institutional Fraud Committee that brought me to trial in 2009 and the explicit lies about Malcolm's fatherly status to my children in July 2009 were just foreplays of a drama that would soon open its curtain in 2009.

The University of Chicago

Institutional Biosafety Committee
5751 S. Woodlawn Avenue
McGiffert House, Room 214, MC 1108
(773)-834-5850

Institutional Biosafety Committee Certification

Approval Date: 08/21/08 8/8/08 sent to NIH-OBA

Investigator: Olaf Schneewind
Office Address: Microbiology
CLSC 1117B

Amendment Number: **AD 04**
Protocol Number: **627**
Protocol Title: Targeting of Yop Proteins by *Yersinia enterocolitica*

Risk Group: RG2
Biosafety Level: BL2
Animal Biosafety Level: ABSL2

Nature of Amendment: **Staff Addition: Staff Additions: Staff Additions: Kristy Skurauskis, Bryan Berube, Claire Cornelius, Nancy Ciletti, Timothy Hermanas, Malcom Casadaban & Laura Satkamp**

Amendment Status: Approved

The amendment to the research protocol described above has been reviewed by the IBC with the results as indicated. Any additional changes to this protocol must be submitted to the IBC and approved prior to initiation. Thank you for your cooperation.

AUG 2008 8/20

Lenora Hallahan
Administrator for Regulatory Compliance

Date

Figure 7F: Malcolm and three other students were certified to be the new staff to Schneewind's Laboratory

Name	Position (Post-doc, technician, student)
Olaf Schneewind	Professor PI
Lauriane Quenee	Research Project Manager
Nancy Ciletti	Senior Research Technician
Derek Elli	Senior Research Technician
Bryan Berube	Junior Research Technician
Melanie Marketon	Visiting professor
Timothy Hermanas	Junior Research Technician
Nathan Miller (RG2/BSL2)	Graduate Student
Bill Blaylock (RG2/BSL2)	Postdoctoral fellow
Will DePaolo	Research associate
Malcolm Casadaban (RG2/BSL2)	Research associate
Dominique Missiakas	PI
Stefan Richter	Research Project Manager
Helene Louvel	Postdoctoral fellow
Anthony Mitchell	Junior Research Technician
Jen Stencel (RG2/BSL2)	Graduate Student

Figure 7G: Malcolm was downgraded from a tenured professor to a Research Associate in Schneewind's Bioterrorism National Lab.

I labored to reconcile all the facts before me and to reflect in particular on what Bill had told us at the end of our discussion, "Malcolm had just been vaccinated." This revelation shook me.

Suddenly, Malcolm cries about poison and refusing made sense; he had talked about his concerns over the safety of the biowarfare products. I believe he objected to the vaccination of laboratory researchers but in the end had no choice.

Vaccine or Poison?

While my thoughts continued to linger on this subject, I quickly surmised the situation in a mental note. *What was in the KIM D27 pathogen that afforded protection to some, but poison to others?*

If I had only had a few more minutes with Bill to ask questions! I thought in frustration. I did not realize, then, that the answer to my question had already been laid out in the last part of Bill's explanation, when he talked about the target delivery of plague bacteria to human blood that determined the septicemic plague death.

I understood then,

- *If KIM D27 was target delivered subcutaneously or intramuscularly to animals and human hosts, it would be taken as vaccine to protect against a plague infection.*

- *If, however, KIM D27 was target delivered to blood, a lethal injection indeed, the result is the fatal septicemic death.*

While I labored over these issues, my children and I were led by Wardenburg to meet with Schneewind who apparently had just finished his PCR experiments and was ready to see us in his office.

Meeting Schneewind Again

Although his office was empty when Wardenburg showed us in and seated us, Schneewind soon appeared, looking exhausted. That was not surprising considering he had a huge problem on his hands. A renowned scientist in Schneewind's GLRCE national lab and a faculty member of the department that Schneewind chaired had died from the plague, from the pathogen Schneewind was entrusted with investigating. Schneewind, I was sure, would have much preferred that the cause of Malcolm's death never be known.

Once Schneewind had sat down across from us, with Juliane standing next to him, I gathered all my courage to begin a dialogue on something I knew best. "An organism with $LD_{50}=10^6$-10^8 should be considered dangerous to animals and humans because a tiny colony on a petri dish, comprising 10^7 bacteria, would be sufficient to kill half of the animals exposed—right?" I asked. I was referring to what Bill had told us about the virulence potential of the KIM D27 (Figures 7A and 7B). Schneewind stared at me blankly, probably taken aback by my expertise on the subject and my knowledge of the virulence potential. I asked again, and I could see he was taken aback by my directness.

I knew, sitting there across from Schneewind, that he had no intention of sharing his conclusions on Malcolm's death with us, much less speculating about any aspect of the situation. I summoned my courage and put my thoughts more bluntly, "Why would such a dangerous organism, the live attenuated plague vaccine, be considered for human vaccination?"

This caught Schneewind completely off-guard. At this point, Wardenburg, who stood next to Schneewind, bent forward to whisper into Schneewind's ear, but her words were loud enough for everyone in that little office to hear.

She said, "Bill has already talked to them about that."

Schneewind nodded slightly, paused, and then said, "I see. Well, there were two other bacterial species in his blood."

Schneewind's answer took me by surprise. Instead of answering my questions, he had thrown a new subject into the maelstrom. It seemed that he was trying to imply that a gram-positive bacteria that was in Malcolm's blood had killed him. I remembered Dr. Howes had talked to us about two gram-positive bacteria that he had diagnosed in Malcolm's pre-mortem blood. I did not know what to make of Schneewind's comment. Clearly Schneewind knew what had caused Malcolm's death and what bacteria were in his blood. But certainly, he did not volunteer any more explanation. My mind whirled. What were these organisms, and how did they get into Malcolm's blood? Had they contributed to his death? How?

By then, Schneewind was evidently in no mood to answer questions and did not feel he had to do so or even remain civil.

"Look, I've spent the entire week working on the memorial for Malcolm. You should give me some respect and credit for all I've done for Malcolm."

Clearly, Schneewind wanted to make us feel guilty about asking more questions about the protocol and practices of his laboratory; worse, he wanted to intimidate us with his supposedly superior professional position and knowledge. I knew that these were the logical questions to ask and that I had both the right and the duty to uncover the details of Malcolm's death. And after all, wasn't that the purpose of our meeting with Schneewind? Who would be in a better position than Schneewind to answer those questions for the family?

A hint of desperation hung about Schneewind that night. He had undoubtedly spent the whole afternoon being grilled by the university administrators and lawyers over his bioterrorism research and its implications in Malcolm's death. All of this was a threat to the future of his laboratory and his professional life. Now the family was asking questions he didn't want asked and didn't want to answer.

Schneewind stood up. "I am not going to sit here and waste my time educating you on something that you wouldn't understand."

That, of course, was a ludicrous statement. He knew I was a published scientist with a doctorate in virology earned from Roizman's laboratory. What I didn't understand was how the plague and other bacteria had invaded Malcolm's system and why no one helped him when he complained of symptoms associated with the plague.

But Schneewind would provide no more information, turning aside any other questions. He said, "I am tired and I have work to do."

With that, we were led out of his office, out of his laboratory, and out of the building. I knew that from this point on, the doors of the University of Chicago would not open to us. Yet on every front—

legally and morally—the university owed the family a truthful explanation of all the circumstances and events surrounding Malcolm's death.

I knew, despite everything, that what we had learned that day, September 18, 2009, was a victory for us. The hidden culprit that brought Malcolm down had finally surfaced. It had a name, a face, an origin, and a time stamp. The organism that had swept through Europe centuries ago known as Black Death had struck again, this time, besieged Malcolm without mercy. Nonrandom in its attack, it was the intended outcome of a plague vaccine that hit Malcolm like a poison dart.

Black Death/Yersinia pestis

During medieval times, a fleet of trading ships regularly navigated the shores of Sicily; then villagers began finding ships adrift, and when they boarded the vessels, they found entire crews dead. The villagers looted the ships for treasure but came away with something else—Black Death. One of the deadliest pandemics to strike the world had begun. Before it burned itself out, millions would be dead.

"*Yersinia pestis* has the dubious honor of being one of the few pathogens that has been responsible for several great pandemics of the disease," according to Richard W. Titball and E. Diane Williamson in their article "*Yersinia Pestis* (Plague) Vaccines," in *Expert Opinion on Biological Therapy*, June 2004: 965–973. "Collectively, these pandemics are estimated to have caused 200 million death worldwide. The most notorious of these pandemics is the so-called 'Black Death', which occurred in many parts of Europe during the 14th–17th Centuries. These pandemic alone are estimated to have killed one-third of the population of Europe."

Now the same organism that devastated Europe had ravaged Malcolm. Though not the same as the wild type plague that infected

Europeans, what besieged Malcolm was a laboratory strain of a live attenuated vaccine strain that had a reduced virulence by 10^7 fold. Since the laboratory strain had never infected anyone in the wild, the devastation was the same, *exitus lethalis*, as in the *Biosafety Manual for Working with Yersinia pestis strain KIM D27 (Figure 5C)*.

The vaccine plague strain had attacked Malcolm in a very specific way that turned the infection into a deadly blow. I would soon find out that the weaponry of the bacteria lies not in the bacteria itself, but how it is directed to cause a human infection that is always fatal.

Foul Play

Wardenburg and Steve Weber from the emergency room called us that evening to confirm that the plague agent found in Malcolm's pre-mortem blood was indeed the KIM D27 strain, the live attenuated plague vaccine strain used in Schneewind's lab. She further stressed that KIM D27 was a Centers of Disease Control (CDC) select agent, and they had to report to CDC and the local public health office for an on-site investigation.

That night, Brooke gathered enough courage to ask Dr. Weber straightforwardly if her father's death had an element of foul play about it. Dr. Weber paused for several seconds, then replied, "Ah! . . . I hope not . . . I don't know . . . not that I know of." Clearly he was not confident in his remarks. But Juliane Wardenburg who was trained in Schneewind's lab and later became a faculty in the Microbiology Department surely would know the detail of experiments that were in progress. But she was silent throughout our questioning.

A Second Autopsy—Dr. Bryant

On September 19, 2009, at 4:00 p.m., dark clouds started to gather, casting the south side funeral parlor where my children and I were gathered into a shadow. We were there to meet with Dr. Bryant, who was hired by the family to perform a second autopsy on Malcolm to resolve many questions we had about the death. After chatting a few moments with my daughters and me, Dr. Bryant said, "I am about to take a quick look at Malcolm downstairs. Anything in particular that you like me to look for during my physical examination?"

I felt a sudden chill down my spine as the weight of this question settled on me. I knew this would be the final chance the family had to find out about the death in our own terms. At this point, my daughters had neither the training nor the emotional capacity to consider the objective medical details. I thought again about what Dr. Howes and Bill Blaylock had told us about the nature of plague infection.

I took a long deep breath. "Dr. Bryant, could you specifically look for external black marks, needle scabs, sores, and wounds on Malcolm's body?"

With that, Dr. Bryant nodded; he knew exactly what I meant. He left us in the room and went downstairs to work.

Thirty minutes later he returned. The three of us were holding our breaths to hear what Dr. Bryant had to tell us. Quickly, he ran through the list of what he had found. He told us then that Malcolm's body was completely free of external black marks or wounds, indicating it was not a bubonic plague infection, and, critically, there was no obvious needle point puncture or scab to support a subcutaneous or intramuscular delivery of the pathogen.

Dr. Bryant's physical examination confirmed for us that Malcolm had suffered a septicemic plague infection, a silent cryptic infection in the blood. However, although the absence of open wounds, sores, eye and lung infections, and scabs eliminated several portals of entries, it meant that the real delivery route was still unknown to us. We could only speculate at the moment. We would have to wait for the autopsy report for real answers.

Dr. Bryant also pointed out that while most of Malcolm's organs, which had been removed from the body during the first autopsy, were left in a bag next to the body, some, including his brain and liver, were not present. That meant Dr. Bryant could not examine them. Dr. Bryant took samples of body and organ tissues to be tested by his autopsy team. Once our autopsy was completed, we had to quickly arrange for cremation because the body had started to decompose.

All of his life, Malcolm had been a simple man who strove to maintain optimal health. He never drank alcohol, coffee, or tea, nor did he smoke or take drugs. He seldom took pain relievers. He worked hard to reduce his weight and control the diabetes he had suffered in recent times. He was almost always in good health. To have his life end with his body and mind ravaged by plague was almost too much to bear.

Taking Malcolm Home

Malcolm's cremation was held on the south side of Chicago on September 21, 2009. It was a beautiful fall morning when Malcolm's parents, John and Dolores, my daughters, Brooke and Leigh, and I arrived at the crematory to say a final good-bye.

I noticed that the sky this day was crispy blue as it had been the winter's day when Malcolm and I had come to the city of Chicago nearly three decades ago to build a family and a life. Young, carefree in spirit, zealous in our love for science, we had driven into the city with confidence and determination. We wanted to do something great, to build a life worthy of ourselves and our children.

Twenty-eight years later, I stood with our daughters, waiting to take Malcolm away from this city that had been the home not of dreams fulfilled but of professional disillusionment and destruction, personal devastation, and death. I felt sick to my stomach, sick at heart, thinking of all Malcolm had suffered at the university, both emotionally and physically. He had been gifted with a genius that only God could bestow, and yet he had remained, always, a gallant man, endearingly naïve, generous, and gentle.

Had God played a joke on him? To have led him to this city, to the University of Chicago, to the Microbiology Department and Department of MGCB, where everything was taken from him, where he had been silenced? Malcolm was alone, so alone when he had been forced into that laboratory of plague and bioterrorism. No one had listened to his cries for help, and no one came to his aid when the plague first overtook him. No colleague stood by his side as he faced the final hours of his life, gasping for every breath.

John Casadaban delivered a prayer with final blessing for his beloved son. Then, together, we took Malcolm home.

8

THE GRAND OLE ELEPHANT DANCE

After the funeral Leigh was hesitant to return to Boston right away. She was only twenty-one, a year shy of graduating from MIT. She had grown accustomed to being with family at times of crisis.

"Leigh, I know you are smart like your Dad. After all, you followed his footsteps to MIT, his *alma mater*." I looked into her teary eyes. "I know your Dad would want you to move on, to continue your education and pursue your dream."

"I know, Mom," Leigh said with some uneasiness. "I was thinking, I wanted to take a year off just to cool things down. But now I see that I'd like nothing better than to be a doctor. I am four years behind. Had I been a doctor already, Dad might have been saved from this horrible infection." She started to sob. "I've always wanted to do this. The medical training will help me with Dad's death."

Leigh boarded the next plane to Boston. At the same time, Brooke and I were on our way home to Southern California. Our trip back was not quite as difficult as when we had come to Chicago. Perhaps we were exhausted from crying so much, from all the heartaches during the last couple of weeks. Instead, I sought new directions and answers that would help resolve remaining questions quickly.

Pandora's Box Opens

The peace of Malcolm's memorial gave way to blistering reports in the national news when the story hit two days later, on September 19, 2009. Pandora's box had opened, releasing pestilence, the mysterious death of a prominent University of Chicago scientist.

Malcolm's death and the implications of a deadly plague pathogen at the University of Chicago took center stage on every TV channel, a story reported every thirty minutes for the next seven days. It pervaded print media and online articles as well, with terms such as *Black Death*, *plague*, *biodefense*, *vaccine*, *poison*, *bioterrorism*, *biological weaponry*, and *public health concerns*.

Malcolm's death had touched everyone in the academic research world; many mourned him and wanted answers. They could not understand how Malcolm, a renowned scientist with four decades of lab experience, could have become infected by a deadly pathogen. Jon Beckwith, Malcolm's early mentor, was devastated by the news and asked Schneewind for the truth in a personal email on 9/14/2009 at 5:16am, 10 hours after Malcolm died (Figure 8A).

> "Dear Olaf.
>
> Thank you for your message. Can you tell me what happened? Although he was a very unusual kind of scientist and person, he was a legend in my lab for his skill, ingenuity and personality. In some ways, his science/"technology" orientation was a forerunner of much of what has fueled the

revolution in molecular biology..." (Emphasis added, Figure 8A)

From: Beckwith, Jonathan Roger [mailto:jon_beckwith@hms.harvard.edu]
Sent: Mon 9/14/2009 5:16 AM
9/14/09

To: Schneewind, Olaf [BSD] - MIC
Subject: RE: Malcolm Casadaban

Dar Olaf,

Thank you for your message. Can you tell me what happened? Although he was a very unusual kind of scientist and person, he was a legend in my lab for his skill, ingenuity and personality. In some ways, his science/"technology" orientation was a forerunner of much of what has fueled the revolution in molecular biology.

My warmest regards to you and Dominique,
Jon

Jonathan Beckwith
Dept. of Microbiology and Molecular Genetics
Harvard Medical School
200 Longwood Ave.
Boston, MA 02115

Figure 8A: Beckwith pled for truth from Schneewind

Beckwith wrote a heartfelt letter to the Casadaban family (Figure 2B), and I read and reread his passages to draw strength as I searched for the truth. "Much of what we do in my lab today, and what other people do in biology, we owe to Malcolm . . . He was in another world in some ways, but what a world he created."

On September 22 in the evening, Kevin Roy of ABC-TV interviewed Brooke and Leigh in a national broadcast about the death of their father. The interview (see http://www.youtube.com/watch?v=rgbMh-J2XZw was conducted in Malcolm's house and his mother, Dolores, was there. In the interview, Brooke and Leigh stressed that the family vowed to do everything possible to seek the truth in the tragic death of their father. Dolores, of course, wanted justice for her son.

As soon as *Yersinia pestis* had been identified in Malcolm's blood, John Burklow, associate director for communications and public liaison at NIH notified its director, Francis Collins. Amy Patterson, NIH's associate director for science policy, forwarded the *Chicago*

Tribune article on Malcolm's death to Allan Shipp, director of outreach for the NIH Office of Biotechnology Activities (OBA), who was "responsible for establishing improved lines of communication with investigators, sponsors, and institutional biosafety committees (IBCs)," and Mary Groesch, senior adviser for science policy at OBA. Amy Patterson remarked, "This will likely get a lot of attention—regardless of what really happened."

From: Burklow, John (NIH/OD) [E]
Sent: Sunday, September 20, 2009 3:11 PM
To: Collins, Francis (NIH/OD) [E]; OD-Small Staff
Cc: Allen, Marin (NIH/OD) [E]
Subject: Fw: Chic Trib: U. of C. researcher dies after exposure to plague bacteria

Figure 8B: email sent to Francis Collins on September 20, 2009 on the death of Malcolm Casadaban

From: Patterson, Amy (NIH/OD) [E]
Sent: Sunday, September 20, 2009 8:17 PM
To: Shipp, Allan (NIH/OD) [E]; Groesch, Mary (NIH/OD) [E]
Subject: IMPORTANT FW: Chic Trib: U. of C. researcher dies after exposure to plague bacteria

This will likely get a lot of attention -- regardless of what really happened --

Figure 8C: email from Amy Patterson to NIH-OBA officers regarding Malcolm's death.

Diminishing Malcolm

On September 20, 2009, one week after Malcolm's death, Schneewind wrote to his Microbiology Department faculty members, staff, and students as well as Roizman and Kanabrocki, in an email:

> "To all Microbiology Faculty, Staff and Students:
>
> As you may be aware from the local news . . . a person who worked in Cummings Life Sciences Center (CLSC601) died last weekend unexpectedly. Yesterday, it was determined that his blood carried an attenuated strain of Yersinia pestis, which may have been the cause of his death. Attenuated strains, such as the kind isolated from the

blood of the deceased, are used for research in our Department and are believed to be harmless to healthy individuals. "

To: 'microbiology-employees@lists.uchicago.edu'; microbiology-seminar-series@lists.uchicago.edu; committeeonmicrobiology@lists.uchicago.edu
Cc: Roizman, Bernard [BSD] - MIC; Knuf, Jamie [BSD] - MIC; Friedman, Meredith [BSD] - MIC; Traw, Emily [BSD] - MIC; Will, Catherine [BSD] - MGC; Quenee, Lauriane [BSD] - MIC; Kanabrocki, Joseph [BSD] - MIC; Easton, John [UCH]
Subject: public health concern
Importance: High

To all Microbiology Faculty, Staff and Students:

As you may be aware from the local news (see the attached documents), a person who worked in Cummings Life Sciences Center (CLSC601) died last weekend unexpectedly. Yesterday, it was determined that his blood carried an attenuated strain of Yersinia pestis, which may have been the cause of his death. Attenuated strains, such as the kind isolated from the blood of the deceased, are used for research in the our Department and are believed to be harmless to healthy individuals.

Appended you will also find a short list of questions and answers about Yersinia and this tragic incident. At this time is seems very unlikely that anyone else is at risk for infection by Yersinia. Nevertheless, laboratory members of the deceased are being offered a prophylactic antibiotic just to be certain. Further, we want to make sure that everyone in our Department is aware of this situation and has all available information.

If you feel you may be at risk or have any health concerns related to this situation please contact Sylvia Garcia-Houchins in infection control via pager (773-753-1880, pager number 8437) or by email (sylvia.garciahouchins@uchospitals.edu).

Please feel free to contact me by email or phone (773 842 2628 mobile) if you have any questions or concerns.

Best regards,
Olaf

Olaf Schneewind
Louis Block Professor & Chair
Department of Microbiology
University of Chicago

Figure 8D: Schneewind announced Malcolm's death to members of his Microbiology Department.

What is clear from this email is that from the beginning Schneewind sought to diminish and depersonalize Malcolm and distance himself from what happened. He referred to Malcolm as "a person who worked in the Cummings Life Science who died unexpectedly." It was as if Schneewind wanted to give the impression that he did not know the "person" well and didn't expect anyone else to recognize the name. Yet Malcolm had been at the university for almost thirty years, had been Schneewind's colleague for more than ten years, and had been a faculty member in the Microbiology Department that Schneewind had chaired for more than seven years.

Malcolm was also a co-investigator on grant AI42797 with Schneewind, which was awarded by NIAID and involved recombinant DNA and *Yersinia pestis* KIM D27. And Malcolm had another grant award, NIH 5-29691, for the development of novel genetic tools for metabolic selection in *Yersinia pestis*, the same pathogen being studied in the National Lab by Schneewind. This renowned scientist and longtime colleague whom Schneewind knew was now reduced to "a person who worked in Cummings Life Sciences Building."

Sadly, this reduction of Malcolm was only the latest of the blows struck against him and his reputation, only the latest methods used to diminish him. In 2005, the MGCB Department had forcefully removed Malcolm's laboratory space and office and moved him to a cubicle shared with retired faculty members in the basement. Malcolm had been forced to use the bench space in Schneewind's laboratory to conduct work on his NIH grants.

Denying Malcolm was not enough. On October 16, 2009, in an Incidence Reporting (a formal NIH document), sent to NIH-OBA, Joseph Kanabrocki, assistant dean of biosafety and associate professor in the Microbiology Department, claimed the following (Figure 8E):

<u>NIH OBA Incident Reporting</u>

"During investigation about this incident, it was learned through interviews with laboratory co-workers that the Deceased, who was employed in the laboratory of Professor Schneewind at a 10% effort, did not always adhere diligently to PPE requirements for BSL2. This behavior was not observed during annual inspections conducted by EHS. "(Figure 8E)

Deviation from IBC-approved containment level or other IBC-approved conditions at the time of the incident/violation: During investigations about this incident, it was learned through interviews with laboratory co-workers that the Deceased, who was employed in the laboratory of Professor Schneewind at a 10% effort, did not always adhere diligently to PPE requirements for BSL2. This behavior was not observed during annual inspections conducted by EHS.

Figure 8E: Incident reporting on Malcolm's death sent to NIH-OBA

Thus Kanabrocki at the university not only depersonalized and diminished Malcolm again, characterizing him as "the Deceased" who was merely "employed" in Schneewind's laboratory "at a 10% effort," but went further, to attack Malcolm's professionalism. He accused Malcolm of careless laboratory practices, of not diligently following personal protective equipment (PPE) requirements for biosafety level 2 (BSL2). (Ironically, biosafety level 2 was the designation given to "work involving agents that pose *moderate* hazards to personnel and the environment.)".

To cover Schneewind and the university, Kanabrocki added they had not reported this alleged recklessness sooner (as required by NIH) because supposedly neither Schneewind nor anyone else in charge had observed this behavior during the annual environmental health and safety (EHS) inspection. So although Malcolm had been working on a grant with Schneewind and in his laboratory for the last four years, neither Schneewind nor anyone else claimed to have noticed Malcolm's failure to "adhere" to safety protocols until after his unexplained death.

This was character assassination, a dagger thrust into the heart of Malcolm Casadaban, who no longer had a voice to defend himself. And how likely was this accusation to be true? Not only was Malcolm a well-respected scientist who had been working with pathogens since the 1970s, but he had also served on the University of Chicago's Institutional Biosafety Committee (IBC) for more than fifteen years during his tenure (Figure 4E). Malcolm, with other members of the IBC, drafted the basic biosafety protocols in accordance with NIH-OBA (Allan Shipp, director of outreach) that governed the biosafety practices day to day in all biological laboratories on campus. His work was embodied in the *Biosafety Manual* as a code to be practiced diligently by all staff, students, and scientists on campus (Figure 5C).

That Schneewind, through Kanabrocki, would attack Malcolm's professionalism in the lab was an outrage. The horror of Malcolm's death was compounded by propaganda generated after his death to destroy his professional integrity and reputation—and to mislead investigators. This meant little to men such as Schneewind and Kanabrocki and all others who spoke falsehoods after Malcolm's death. In so doing, they also put the University of Chicago's reputation at risk, an embarrassment to the hood and gown they wore that represented the university.

Malcolm had been alone, all alone, in this world of academic arrogance and conflict. He had run the south side streets of Chicago in the middle of the night to clear his head. He had borne the mockery from his colleagues in the department and stood by helplessly as the unversity, through a bogus fraud investigation, targeted me. He had grieved in his private corner, feeling he had nowhere to turn. The wretched fire burning inside of him finally had burst out on July 15, 2009, and his cries had shown exactly how he felt about his life, the fear deep inside his soul. He had felt alone and lost in the academic fortress that he was so accustomed to and he feared what the future would bring.

Moffat had assured Malcolm that nothing would happen to him, that his tenure appointment would not be at risk. Though Malcolm found comfort in what Moffat had told him at the time, Moffat kept Malcolm in Schneewind's laboratory and did not protect him from the mysterious attack that had taken his life. (Moffat was the principal investigator of the Howard Taylor Ricketts Biocontainment Laboratory at the University of Chicago.) No one there had. They stood on the sidelines watching one of their colleagues disparaged, belittled, and finally brought down by the ill forces that hovered over him. And now they kept their silence about what had really happened to Malcolm.

Was not an American President, John F. Kennedy, who said, "The only thing necessary for the triumph of evil is that good men do nothing." I repeatedly asked God where and from whom could Malcolm find justice?

After I left Chicago, rage grew in my heart as the circumstaces and facts about Malcolm's death gradually poured in. One night when I was in bed, Malcolm came into my room and sat down across from me, under a dim light. His back was towards me, his face hidden in the dark. Still, he spoke in his usual soft voice, and said calmly, "Vaccines they gave me . . . Poison . . . Poison . . ."

"Vaccines They Gave Me... Poison... Poison..."

The word *vaccines* crept echoes in my mind. I jumped up quickly. When I turned around to look for him, he was gone. It had dawned on me that these were the precise words recorded in the FDA's Center for Biologics Evaluation and Research (CBER), the same words that marked Malcolm's death by poison. In addition, as I was told, the research on bioterrorism was a joint project overseen by the DOD, NIH, and the FDA. Malcolm died from KIM D27 vaccination, what he received was a poison dart that sent him to his death.

Suddenly it all made sense. The senior staff in Schneewind's laboratory told the family that Malcolm had just been vaccinated. Dr. Howes told us that Malcolm suffered a fatal septicemic illness. The vision that lingered in my mind was that of a deadly pathogen, a poison, a weapon of bioterrorism, a missile of death that had been locked on Malcolm then released to strike him down.

When I asked Schneewind on the night of September 18, 2009, about the safety of a live attenuated vaccine that he administered to his students and staff, he threw me out of his office. Schneewind's violent reaction to my theory was my first evidence that I had correctly surmised how Malcolm had died, and how Schneewind's laboratory staff was being used as human guinea pigs in plague experiments.

Vaccine Trials—Human Trials

University of Chicago officials told the Chicago Breaking News on September 19, 2009, (*http://articles.chicagobreakingnews.com/2009-09-19/news/2850488319/news/28504883_1_death-notice-plague-researcher).*

> "[T]the weakened strain of the bacteria was used as a vaccine to protect against the plague"

It was also bioterrorism warfare in development in Schneewind's laboratory.

I needed more information. I remembered Stuart Shapira, a fine young man and a former student of Malcolm in early 1980. He was now a medical officer and acting associate director for Science at National Center on Birth Defects and Developmental Disability of CDC (Figure 8F). I sent him an email asking if he knew anything about Malcolm's death. He answered on October 5, 2009, just three weeks after Malcolm died.

> "[T]the strain of Y. pestis that was isolated from Malcolm is the same as the attenuated strain in the lab (strain KIM D27)." (Figure 8F)

He continued,

"This strain is the one that has been used in vaccine studies" (Figure 8F).

From: Shapira, Stuart (CDC/CCHP/NCBDDD) (cso6@cdc.gov)
To: plumfairy@sbcglobal.net;
Date: Mon, October 5, 2009 10:25:49 AM
Cc:
Subject: RE: Hi, This is Joany

Hi Joany,

I apologize for the delay in responding to your e-mail, but I was out of the office for a couple of days and it took me a while to determine who is working on these issues. I found out that the work is actually not going on at the CDC site in Atlanta, but rather in a laboratory of the Division of Vector-borne Infectious Diseases at a CDC site in Ft. Collins, Colorado. I spoke with individuals in Colorado and obtained the following information:

The very preliminary testing indicates that the strain of Y. pestis that was isolated from Malcolm is the same as the attenuated strain in the lab (strain KIMD27; also known as KIMD5). This strain is referred to as "pigmentation negative" because it has a 102 kb deletion that comprises the high-pathogenicity island (HPI) involved in bacterial iron uptake, as well as the pigmentation segment (pgm) associated with Congo red staining of colonies. This strain is the one that has been used in vaccine studies. Several references for this strain are as follows:
--Stuart

Stuart K. Shapira, M.D., Ph.D.
Medical Officer
Acting, Associate Director for Science
National Center on Birth Defects and Developmental Disabilities (NCBDDD)
Centers for Disease Control and Prevention

Figure 8F: email from Stuart Shapira noting KIM D27 recovered from Malcolm pre-mortem blood was a strain used in vaccine studies.

On October 5, just weeks after Malcolm's death, I had confirmation (Figure 8F). The strain isolated from Malcolm's pre-mortem blood was KIM D27 and KIM D27 was the strain used in human vaccine trials in Schneewind's laboratory. I vividly remembered the *Biosafety Manual for Working with Yersinia pestis strain KIM D27* on Malcolm's desk in Chicago that noted *"exitus lethalis"* (Figure 5C). My immediate reaction was an unbelievable horror that sent chills down my spine. No wonder Schneewind had freaked out when I questioned him about the virulence of KIM D27 in a human trial. I became charged as I continue to seek the truth and nothing but the truth.

Threats Against Malcolm's Daughters

Although reluctant to do so, given her earlier conduct, I turned to Ling Chan in the university's Department of Biochemistry as I continued my investigations. Whatever the reasons behind her false accusations to Malcolm, I had known her since college and we had been colleagues at the university. I needed information and thought she could provide it. I was unprepared for what followed when I called her and started asking questions.

"If you or your family ever decide to file a lawsuit against the university," Ling said, "the university will do anything and everything to retaliate against your children." She didn't bother to threaten me, knowing that threatening my children would be much more frightening.

"They would suffer irrevocable damage to their lives and their careers just as Malcolm did." Ling continued. "Leigh is only twenty-one. She is so young and has her whole life ahead of her. Brooke too."

Shocked and outraged, I could only listen. I could not be sure if Ling meant that my children's mortal lives were at risk or just their general well being. Ling was clearly angry at the prospect that the family would consider a lawsuit, but the idea of a lawsuit had not even entered our minds yet. What moved Ling to become so angry, to have jumped the gun and made a preemptive strike against my children in response to a lawsuit that did not exist?

"The university is not short of money," Ling continued to rant. "In fact, the university is very well funded. But we are not going to make any payment to Malcolm's family. And we are not going to allow the university's reputation to be damaged in a lawsuit. You know what I am talking about. I am speaking the words from You-Know-Who."

I couldn't believe what I was hearing. Shortly after the death of a faculty member, the University of Chicago and its cohorts showed no compassion for the dead or his family. Instead, they verbally attacked and abused the family, as if whatever anger they harbored toward us had not subsided with the death. I wondered what Malcolm had done to the university, the faculties, and his colleagues, in their minds, to justify such action, let alone the fact that Malcolm's death might have been brought about by the research programs of the GLRCE. It was difficult to deal with my grief as the facts started to pour in.

When I thought about all that had happened to me, to Malcolm, to our family since my successful lawsuit against the university—the verbal harassment against Malcolm at the university, Malcolm's loss of his lab and other demotions, the abortive university fraud investigation against me, the slander about a supposed affair—they all seemed like gangland acts of retaliation. I got a sense from Ling that Schneewind, Roizman, Mahowald, Moffat, and Ling herself considered themselves an elite cohort, their actions above the law.

I couldn't speak, I was so overwhelmed, and into my silence, Ling said, "Malcolm is dead. Dead man cannot talk. We have to help those who are still living to move forward over this tragedy."

That statement burned away some of the shock, and I asked her, "If it was your husband who was killed, what would you do"?

"We are not talking about my husband. *My* family is not the issue! Don't start with my husband," Ling said, anger blazing in her voice.

Her words placed Malcolm, as well as our daughters, in a subordinate class to those completely loyal to the university, who supported and served its purposes. She also painted a chilling picture of Schneewind, portraying him as a military commander with absolute authority and power on every matter and every person in his lab and in the Microbiology Department that he chaired. Such chilling thoughts would haunt me in the coming days, weeks, and months. Could this be a reality in America, the country to which my family and I had come fifty years earlier seeking freedom and justice? Where were those values now for Malcolm and my family?

Before I could hang up on her, Ling asked for Casia's whereabouts, indicating that she would like to speak to Casia about what she just said to me. *More threats?* I wondered.

After I hung up, I noticed my hands were shaking. She had scared me; I admitted it. If I insisted on investigating Malcolm's death, would resources be brought against my daughters? Would the university attack me personally and professionally, destroy me as Malcolm had been destroyed?

But then the rage began to build. What had really happened to Malcolm behind the closed doors at the University of Chicago? I wanted the truth! I needed more evidence to back up my theory. I knew Malcolm had been poisoned, poisoned by KIM D27. It looked more and more as if he had been forced into a human trial.

Federal and State Investigators

Although my anger was strong, I wasn't sure how to go about getting answers. So I was greatly relieved when I learned, on September 21, 2009, that a team of federal and state investigators had arrived on campus to investigate the plague death. Finally, we were going to find out the truth about Malcolm. I wanted to believe that these experts would find the real causes and shed light on Malcolm's death.

Kathy Ritger, M.D., M.P.H., from the Chicago Department of Public Health, contacted us during this investigation. She led the team of investigators. I thought well of Kathy. She was a kind individual who expressed genuine sadness for Malcolm's loss. In our first conversation, she introduced herself and told us that her position was a neutral one, that she was only interested in discovering the facts surrounding Malcolm's death. She promised to call us every couple of days to keep us updated on what she had discovered and would ask us for our help in her investigation as needed. I thought she would be that angel who would seek justice for Malcolm in the days ahead.

Members of the federal and state Investigation teams and the university team were as follows:

Federal Team:

Martin Schriefer, PhD, Chief, Diagnostic and Reference Laboratory, CDC Fort Collins, CO
Kristen Metzger, MPH, CDC/CSTE, Chicago Department of Public Health, Chicago, IL
Stephanie Black, MD, MSc, Medical Director, Chicago Dept of Public Health, Chicago, IL
Salvatore Cali, MPH, CIH, Hygienist, UIC, Great Lake OSHA, UIC School of Public Health, Chicago, IL
Sue Gerber, MD, Chicago Public Health, Chicago, IL
Brad King, MPH, CIH, Hygienist, NIOSH, CDC, Cincinnati OH
Paul Mead, MD, MPH, Chief, Epidemiology and Surveillance, National Center, CDC, Fort Collins, CO
Andrew Medina-Marino, PhD, IL Dept of Health, Chicago, IL
Kathy Ritger, MD, MPH, Chicago Dept of Public Health, Chicago, IL
Kingsley Weaver, MPH, Chicago, Dept of Public Health, Chicago, IL
Chicago POLICE, Sheriff and law Reinforcement Officers.

University Team:

Caroline Wilson, Chief Operating Officer & Associate Dean
Jane Schumaker, Associate Dean for Administration
Dr. Joseph Kanabrocki, Dean of BioSafety Officer
Krista Curell, Risk and Patient Safety
Olaf Schneewind, Microbiology; Investigator, Chair of Microbiology

Dr. Karen Frank, Pathology
Judd Johnson, Facilities
Bill Huffman, Facilities
Dr. Steven Lelyveld, Occupational Medicine
Caroline Guenette, Occupational Medicine
John Easton, Communication and Marketing and/or Julie Peterson
William Frazier, Legal
Dr. Steven Weber, Hospital Epidemiologist and Infection Control
John Satalic, Legal
George Langan, Director Animal Resources Center
Dr. Eric Whitaker, Executive Vice President

9

"UNTREATED ... SEPTICEMIC ARE FATAL"

Once I got back home after the turbulent weeks in Chicago, I felt both triumphant and sad. An army of federal and state investigators had flooded to the university to investigate the death. I assured myself that I would soon see the light.

Despite the fact that there was an official investigation in progress, I started my own investigation in a quiet corner away from the center of the quake. I began to organize my thoughts, to assemble the pieces of puzzles that had come out of my recent inquiries. No matter how incomprehensible as they appeared then, I knew something would come out of it. I made a list of questions:

- What is the true nature of KIM D27?
- What was in that witch's brew that had cost Malcolm's life?
- And how was it delivered?

Departmental Announcement of the Death

On September 20, 2009, Schneewind sent an email to his Microbiology faculty, staff, and students in which he said, "As you may be aware from the local news . . . , a person who worked in Cummings Life Science Center (CLSC601) died last weekend unexpectedly. Yesterday, it was determined that his blood carried an attenuated strain of Yersinia pestis, which may have been the cause of his death" (see Figure 8D).

Schneewind had carefully hidden Malcolm's identity, referring him as a "person" to his department. I was disheartened by his lack of concern for Malcolm, who had been a friend, a colleague, and a member of Scheewind's Microbiology Department for many years (Figure 8D). It left a bad taste in my mouth, but I kept my suspicion to myself at this point.

FAQ #2: CLSC--Y. PESTIS UPDATE, SEPT. 25

What do we know about the researcher's cause of death?
We may never know exactly what caused his death, but a thorough investigation of his laboratory work and potential exposure is underway, with multiple experts looking at many different aspects. Given the epidemiology to

Figure 9A: "We may never know exactly what caused his death"

The True Nature of KIM D27: Safe and Harmless?

In another Q & A document from the University of Chicago, Schneewind said,

> "[T]he weakened strain of *Y. pestis* lacks the bacteria's harmful components. This strain is not known to cause

illness in healthy adults and has been used in some countries as a live-attenuated vaccine to protect against plague." He continued, "The weakened strain is not believed to be dangerous to healthy individuals." (Figure 9B).

It is a weakened or "laboratory" strain of Y. pestis that lacks the bacteria's harmful components. This strain is not known to cause illness in healthy adults and has been used in some countries as a live-attenuated vaccine to protect against plague. It has been approved by the Centers for Disease Control and Prevention (CDC) for routine laboratory studies. The weakened strain does not require the special safety precautions required for work with virulent strains.

If it is not dangerous, why did he die?

Testing is underway, and the Medical Center is working with public health officials to find out all that can be known about this case. The weakened strain is not believed to be dangerous to healthy individuals, but underlying health conditions could increase susceptibility.

Figure 9B: Schneewind said, the weakened strain of Y. pestis is safe and harmless, not dangerous to health

Really? I was puzzled. "KIM D27 is not known to cause illness in healthy adults"? Yet I vividly remembered the *Biosafety Manual for Working with Yersinia pestis strain KIM D27*, which warned about the symptoms of plague infection and "*exitus lethalis*" (see Figure 5C). And why did Malcolm die? I felt disheartened by the conflicting statements made by the man who was a M.D., Ph.D., the principle investigator of the GLRCE, the Louis Block Professor and the chair of the Microbiology Department. Whose statements could be trusted?

I wasn't the only one asking that question. In the September 22, 2009 edition of the *New Scientist*, Debora MacKenzie questioned another University of Chicago statement about the safety of the plague strain Malcolm had been working with (Figure 9C).

> "The University of Chicago has not revealed which strain of the Yersinia was, but says it should have been safe,"
>
> > "This strain is not known to cause illness in healthy adults and has been used in some countries as a live-attenuated vaccine to protect against plague. It has been approved by the Center for Disease Control and Prevention (CDC) for routine laboratory studies. The weakened strain does not require the special safety precautions required for work with virulent strain." [Quote from Schneewind in Q&A (Figure 9B)]
>
> "But the case reminds me of misgivings that have been voiced about weakened strains, which might not be as safe as they seem. Worse, up to a fifth of people vaccinated with EV76 develop flu symptoms such as fever, headache, weakness and malaise."
>
> "The strain seems likely to have been EV76, which is used as a vaccine in Russia and Madagascar. It is considered unlikely to revert to the virulent strain. But even without reverting, it kills some mice immunized with it." (Emphasis added, Figure 9C)

Wrote MacKenzie.

One of her most telling remarks was the simple question,

> "So what's going on?"

Mystery surrounds 'plague death' victim

18.05 22 September 2009

Health Science In Society

Debora MacKenzie, consultant

Casadaban was reported to have been working with a weakened vaccine strain of *Yersinia pestis*, intended for the development of vaccines against plague. However doctors are not sure the bacilli in Casadaban's blood were the cause of death or how they got there. So what's going on?

Y. pestis has been linked to the medieval Black Death, though proof has been elusive. We do know that it kills several thousand a year by causing bubonic plague, and it is considered a potential biological weapon.

The University of Chicago has not revealed which strain the *Yersinia* was, but says it should have been safe:

> This strain is not known to cause illness in healthy adults and has been used in some countries as a live-attenuated vaccine to protect against plague. It has been approved by the Centers for Disease Control and Prevention (CDC) for routine laboratory studies. The weakened strain does not require the special safety precautions required for work with virulent strains.

But the case reminds me of misgivings that have been voiced about weakened strains, which might not be as safe as they seem.

The strain seems likely to have been EV76, which is used as a vaccine in Russia and Madagascar. It is considered unlikely to revert to the virulent strain - but even without reverting, it kills some mice immunised with it.

Worse, up to a fifth of people vaccinated with EV76 develop flu symptoms such as fever, headache, weakness and malaise, according to Rick Titball of the University of Exeter, UK, a leading *Yersinia* expert. Some even need hospitalisation, he says, though no deaths have ever been reported that we know of.

Of course, we don't yet know whether these findings have any bearing on Casadaban's tragic death. For that we'll need to which strain was in his blood and whether it was the bacteria that killed him.

Research on *Yersinia* increased in the US after 2001, prompted by increased investment in research into potential bio-weapons following the anthrax attacks.

Figure 9C: Debora MacKenzie responded to Schneewind's comments on the safety of the plague strain

Freedom of Information Act (FOIA)

I was sitting at my desk, wondering what avenues I could pursue to learn the truth, when a memory flashed through my mind. The memory came from a time when Malcolm and I were still married, still living in Chicago, before everything had started going wrong. A friend of Malcolm's, a fellow faculty member, had come over for the evening. Smiling broadly, obviously delighted to have this visitor in our home, Malcolm introduced us. Then he and his visitor headed down to the basement, which we had set up as a family room. They were already deep in discussion before the door swung shut behind them.

The remembered slam woke me from my trance. I thought of this friend of Malcolm's who had become a friend of mine as well. Although in Europe at the time, he had sent flowers to the family when he heard about Malcolm. I wasn't sure why the memory of that long-ago day had come to me, but I decided it was a sign that I should call him.

When he answered the phone, he said, "By the Freedom of Information Act, I request to know what happened to Malcolm and how he died."

The words were so unexpected that I nearly laughed—a moment of gallows humor, as they say, for both of us. We talked long after that, about Malcolm and the present events as well as about others times, our conversation punctuated by choked pauses and tears.

Only after I ended the call did the real power of the phrase "Freedom of Information Act" strike me. It was like beacon in the night, a bright key to opening locked doors. But it only applied to federal agencies. Malcolm's death was a local event, wasn't it, concerning a state and a university? Would federal agencies, such as NIH, have files about his death? On the other hand, he had been infected by a plague pathogen being researched at GLRCE and a Regional Biocontainment Laboratory, established by the federal government and paid for by U.S. taxpayers. I thought it was at least worth a try.

The next day I filed my FOIA request with NIH, the agency I felt would be most likely to have information about Malcolm's death. Within two days of the filing, an NIH officer had called me

"Dr. Chou, we want to help you."

I had not mentioned Malcolm's name in my FOIA request but provided minimum information about a plague death at the University of Chicago. But this NIH officer clearly knew who I was and what information I was seeking.

"I am from the front office at NIH," she said, making sure I knew her official position, "and I want to help you."

In those days of grief and uncertainty, when I stood alone trying to find the truth about Malcolm's fate, anyone who offered kind words immediately brought out my vulnerability and gratitude. "Thank you," I said and waited to hear more.

"You need to modify your FOIA request to include the name of the victim," she continued. "Dr. Malcolm Casadaban, right? . . . Casadaban?"

She knew; she could help! Relief flooded me.

"You need to add to your FOIA request to include emails and documents regarding the death event and the victim. When the information becomes available, we will do our best to get it to you."

These were the words that I was desperate to hear. At last, someone had given me hope again. More importantly, from her response, I knew NIH must have, or expected to have, documents to share.

Speaking to an NIH officer was like talking to an old friend; I often consulted with NIH officers during my academic research career. I told her that I was once a graduate student supported by an NIH training grant and a postdoctoral under an NIH fellowship. The National Cancer Institute had awarded me a grant for later research.

I needed her help and help from NIH in my quest for truth. She listened with a sympathetic ear about my endeavors. I opened up, telling her of my doubts and questions about KIM D27 being used as a vaccine in human trials. Knowing how the NIH grant review process worked, I asked her, "How did a human trial employing deadly pathogens ever get approved by the NIH Review Board?"

Even though she was silent for the most part, I felt comforted and relieved. There was sympathy in the words she did offer and in her voice, which touched my heart. She then kindly directed me to several other federal agencies that she knew would have the information I so needed. Malcolm's name was clearly familiar to the federal agencies involved in the projects dealing with bioterrorism.

Attenuated Strains: EV76 and Human Trials

While I waited for the FOIA material, I began reading a study on another attenuated strain of *Y. pestis*, EV76, which was used in human trials conducted in the former Soviet Union (FSU) prior to 2004. Debora MacKenzie had referenced EV76 and a study by Richard Titball in her article. What made reading about EV76 and the trials so compelling for me were the parallels with KIM D27, the culprit in Malcolm's case.

Richard W. Titball and E. Diane Williamson wrote in their 2004 article "*Yersinia Pestis* (Plague) Vaccines" (*Expert Opinion on Biological*

Therapy 4[6]: 965–973, http://informahealthcare.com/doi/abs/10.1517/14712598.4.6.965):

> "There has been concern that *Y. pestis* might be used illegitimately as a bioterrorism or biological warfare agent." (Emphasis added).

Debora MacKenzie noted also,

> "Research on Yersinia increased in the US after 2001, prompted by increased investment in research into potential bio-weapons following the anthrax attacks." (Emphasis added, Figure 9C).

And Schneewind noted in his GLRCE mission statement and in Bioterrorism Protocol, part III.

> "Yersinia pestis, the highly virulent agent of plague, is a biological weapon." (Emphasis added, Figure 5A)
>
> "Yersinia pestis remains a major threat for bioterrorism and biological warfare." (Emphasis added, Figure 10J)

A section of this article addressed the use of EV76 in human trials in the former Soviet Union and in the former French colonies (and still being used in those territories). EV76 is a natural isolate of a live attenuated plague vaccine strain, similar in genomic structure to KIM D27 and in pathogenic profile, sharing extensive homology.

Titball and Williamson discussed the animal model system in which EV76 was delivered subcutaneously. "There are significant concerns over the safety of this vaccine since vaccination of mice or vervets with this strain can result in fatalities" (Figure 9E).

I recalled the danger of KIM D27, similar to EV76, had been marked with the end point of infection in "*exitus lethalis*" in the *Biosafety Manual for Working with Yersinia pestis strain KIM D27* from Schneewind's laboratory. Titball noted the danger levels of EV76 in human trial in subcutaneous delivery: "studies in the FSU have reported that some vaccinees (1–20% depending on the route of vaccine administration) develop a febrile response with headache, weakness and general malaise and individuals with severe reactions required hospitalization" (Figure 9E). Noteworthy was the subcutaneous delivery of EV76 in the human trials. Septicemic delivery involving EV76 was not reported here, possibly because any other delivery other than intravenous was not available before 2004. I had the gut feeling it would be like a lethal injection of 100 percent fatality, as *"exitus lethalis"* for KIM D27.

> inhalation challenges [25]. However, there are significant concerns over the safety of this vaccine. The vaccination of mice or vervets with this strain can result in fatalities [25,26]. Of greater concern, studies in the FSU have reported that some vaccinees (1 – 20% depending on the route of vaccine administration) develop a febrile response with headache, weakness and general malaise and individuals with severe reactions required hospitalisation [26,27].

Figure 9D: Human trial of vaccination in Russia (FSU)

Animal trials with EV76 can result in fatalities. And yet EV76 had been administered subcutaneously to humans in the vaccination trials in FSU. I felt horrified by Titball and Williamson's report when I thought about the similarities between EV76 and KIM D27. That made the fact that KIM D27 was being used to inoculate laboratory researchers in the United States—in essence human trials—in 2009 even more frightening. Could it have gone further still? Could someone have used the required vaccine to target Malcolm Casadaban? The thought was horrifying, but no one had yet explained how Malcolm had become infected.

I remembered well what Dr. Howes had taught us that night when the family gathered to meet with him to learn of Malcolm's final hour of passing at the emergency room. As Dr. Howes artfully put it, "Human blood is a sanctity pool. It is the holy water that runs in our body to sustain our survival. If one bacteria, a blood borne pathogen, ever successfully reached blood, it could set off a full-blown septicemic infection with 100 percent fatality rate in just a few days."

So devastated was I at the idea of what Malcolm had suffered that for several days, I could not get out of bed. Nightmares of all kinds invaded my mind, my senses, awake or sleeping.

Two-Day Infection Protocol

September 2009

		1	2	3	4	5
6	7	8	9	10	11	12
13	14	15	16	17	18	19
20	21	22	23	24	25	26
27	28	29	30			

A call from Brooke roused me from my depression. She had the investigator Kathy Ritger on the other end of the call; Kathy had called to update us on the findings of the official investigative team. First, she noted a report by a hygienist on the team that a filter to one of the hoods in Schneewind's laboratory had been newly replaced prior to the team's arrival. (Kathy thought this might be relevant in Malcolm's death, an inhaling exposure, but later found this was not how Malcolm had been infected.)

In another issue, Kathy said that the investigative team felt the need to sequence the entire genome of the KIM D27 isolate recovered from Malcolm's pre-mortem blood. The goal was to determine if any of the harmful components had found their way back to the genome to cause the deadly infection. Despite Schneewind's attempt to do the sequencing on site, the investigation team preferred to use a third-party facility at Northern Arizona University to avoid the apparent conflict of interest. Her words gave us the confidence of a fair and unbiased, fact-driven investigation that would soon deliver the truth. She would surely bring justice to Malcolm. The truth would prevail

Kathy then told us something that baffled me. She said, "Malcolm got infected in two days, the first was on August 31 and the second being on September 1 of 2009." She paused then, as if checking her notes to affirm the accuracy of her statement. She then confirmed that it was on September 1, 2009 that Malcolm actually got infected by KIM D27.

"How strange! A 'two-day' infection," I murmured to myself. If Malcolm's infection was an accident as Schneewind claimed in his Q & A on September 25, 2009 (Figure 9A), how could she pin the infection to the exact date and what was the "two-day" infection about? The infection sounded like a medical regime of a protocol rather than a random accident. I could not immediately respond to her words, but they were engraved in my mind now and the problem they presented would continue to arise each time I asked how Malcolm got infected in the first place.

Whatever the truth, I was now certain that there would be no simple answer to the cause of Malcolm's death. Perhaps the investigators knew more than they were saying about what exactly was involved in the "two-day infection" protocol—more, what had been in the witch's brew that brought Malcolm down. Days later, when I reflected upon what Kathy meant by the two-day infection, I was almost certain that she knew or suspected information that she was not sharing with the family. I was also growing ever more certain that Malcolm's death had not been random at all. The suggestion of foul play was in the air, but no one dared to utter those words.

FOIA Documents Arrive

Within weeks, the first set of my FOIA documents arrived. I went through them frantically to find those that precisely described the virulence of KIM D27. My theories would soon find facts supporting my belief that Malcolm's death was not at all random, that it had been a calculated and a highly orchestrated event under the guise of lawful bioterrorism research and development.

The case took a big turn when Schneewind's pathogenic profile of KIM D27 sent to NIH surfaced. It was the set of documents submitted to NIH for approval before what were, essentially, human trials began in Chicago.

I soon noticed from the pattern of the protocols that National Labs, by NIH policies, were to submit their Protocol Submission and Agent Profile statements to NIH for progress review on a semi-annual basis. (The passages cited in this Chapter and in later Chapters were excerpts taken from those documents.) These were prepared by Schneewind, the PI of the bioterrorism laboratory, sent to the university IBC select agent Committee for approval before submitting to NIH for review. (I have provided these documents in the Appendix of this book for the reader's benefit.)

Among these documents, one immediately caught my eye. The pathogenic Profile of KIM D27 submitted by Schneewind to the IBC Select Agent Committee on October 7, 2008; approved by Committee Chairman David Pitrak on the 14th before being sent to NIH for review. This was eleven months before Malcolm's death.

The Code of Death: "Untreated ... Septicemic are Fatal"

In Section I of KIM D27 *Pathogenicity*, Schneewind summarized the virulence potential from the animal studies he had conducted earlier. Again the route of delivery determines the disease outcome.

- Untreated bubonic plagues (50 percent fatality rate)
- Untreated pneumonic plagues are fatal (100 % fatality rate)
- Untreated septicemic plagues are fatal (100 % fatality rate)

The code of death that struck Malcolm was embodied in the language like an inscription in the first page of the agent profile forms of KIM D27: "Untreated pneumonic and septicemic [plagues] are fatal" (Figure 9E, Appendix 3: Pathogenic Profile of KIM D27). These four underlined words provided the backdrop of Malcolm's death. Malcolm suffered a KIM D27 plague infection to his blood, which was left untreated, and he died of a septicemic infection. It looked more and more like the infection and lack of treatment were intentional.

Schneewind, Alexander, and Kanabrocki, infectious disease experts and members of the Microbiology Department of the university had been saying that this strain of the pathogen was safe and harmless (Figure 9A, Q&A). Yet to NIH, Schneewind described KIM D27 pathogenesis as "untreated septicemic are fatal," and the IBC *Biosafety Manual* warned that "*exitus lethalis*" could result from infection by KIM D27.

PATHOGENICITY: Zoonotic disease; bubonic plague with lymphadenitis in nodes receiving drainage from site of flea bite, occuring in lymph nodes and inguinal areas, fever, 50% case fatality if untreated; may progress to septicemic plague with dissemination by blood to meninges; secondary pneumonic plague with pneumonia, mediastinitis, and pleural effusion; untreated pneumonic and septicemic are fatal

Figure 9E: Select Agent of KIM D27 strain (RG2) and its pathogenicity

To my amazement, the statement, "Untreated septicemic are fatal" did not specify conditions to iron dependency for KIM D27, a condition that had been attributed to Malcolm's iron overload and his death. I understood that the blood environment of a normal individual has at least twenty-two metabolic pathways to produce iron. That in itself confers the ability of KIM D27 to multiply proficiently in the blood. Genetic disposition of iron overload in a hemochromatosis individual does not enhance pathogenesis in the KIM D27 infection in blood. Thus, in Schneewind's mind when he drafted the KIM D27 pathogenesis, iron load is not a culprit in

Malcolm's death. (This clue will be discussed in Chapters 10 and 11 in relation to Malcolm's autopsy report).

Vaccine or Poison?

In the second page of the same document, it noted an immunization program in place in the bioterrorism/biocontainment laboratory for the human trial (Figure 9F).

> "Immunization is recommended for personnel working regularly with culture of *Y. pestis* or infected rodents, and boosters are required every 6 months if high risk continues." (Figure 9F)

IMMUNIZATION: Although field trials have not been conducted to determine the efficacy of licensed vaccines, experience has been favourable; immunization is recommended for personnel working regularly with culture of /*Y. pestis*/ or infected rodents, boosters are required every 6 months if high risk continues; protection against pneumonic form is limited

Figure 9F: Immunization protocol and vaccine boosters every 6 months are for human trials in Schneewind's Laboratory

In another document, there appeared an IRB (Institutional Review Board) Protocol, 15672A, that dealt specifically with human subjects and their biological specimen collected in human experimentation (Figure 9G). It would appear that human trial by KIM D27 had been approved by IBC, later by NIH in 2008, prior to Malcolm's death in 2009.

e. ☒ Human Subjects (including biological specimens*)

IRB Protocol #	Approval Date	Gene Transfer Study?**
15672A	11/1/07	Yes ☐ No ☒
		Yes ☐ No ☐
		Yes ☐ No ☐

* Please contact the IRB for more information 702-6505.

** Requires the PI to ensure that all aspects of Appendix M of the NIH Guidelines have been addressed. For complete requirements, please review Section IV-B-7-b-(6) of the NIH Guidelines and contact the IBC Office.

Figure 9G: IRB Protocol, #15672A that specifies conditions of human trials had been approved by NIH on 11/1/07, ready to take place in Schneewind's laboratory

These paragraphs clearly revealed a laboratory protocol to vaccinate healthy adults with KIM D27, a protocol conducted in a private, closed-door human trial on the campus of the University of Chicago. Shapira's email on October 5, 2009, also supported the notion that the strain isolate recovered from Malcolm, KIM D27, was the strain that was used in the human vaccine trial in Schneewind's laboratory (Figure 8F). And the hapless candidates of the vaccine trials were the researchers at Schneewind's laboratory, unknowing, unwilling participants in a laboratory game of Russian roulette (Figure 9H). Malcolm was one of them, demoted from a tenured professor to a research associate to a laboratory worker in Schneewind's laboratory. Malcolm at age sixty, a tenured professor, should not have had to submit himself for the vaccination by a potentially deadly pathogen.

This private immunization was not in compliance with the federal clinical trials guideline registered with FDA as required of all human trials, making it the greatest example of unethical U.S. human experimentation in the twenty-first century.

It all became clear to me that there was a biological weapon in development—a secret weapon that destroyed human lives.

A Pandora box had opened. These facts and inferences fit my theory of Malcolm's death, and his mysterious July 15 rant, "DANGEROUS . . . REFUSED . . . DEADLY," suddenly took on a whole new meaning with new resolve: the plague vaccine experiments on the healthy adult researchers was too "dangerous," and Malcolm "refused" to take part because he knew it was "deadly." I recalled the mandatory vaccination protocol in Schneewind's laboratory that had been in place since 2004 that required laboratory researchers to be vaccinated in order to keep their jobs in the bioterrorism laboratory at the University of Chicago (Figure 7E).

It was the dual function of the KIM D27 that could be used as vaccine to protect against plague infection as well as weaponry to cause fatal septicemic death. It began clear to me that the weaponry of KIM D27 lies not in the bacteria itself but in how it was delivered to cause a fatal infection. The question that haunted me for days at this point was how the bacteria gained entry into Malcolm and caused a septicemic infection that devastated his life.

The revelations and conclusions from these federal documents threw me into a painful chaos realizing how the bioterrorism weaponry could be used to harm Americans in the land of the free, this model country for constitutional freedom.

It was a mortal sin to experimentally vaccinate a group of scientists whose own goals were to study science to benefit the human race. At the University of Chicago, researchers and faculties had become the human guinea pigs to advance the development of bioterrorism weaponry. These thoughts horrified me, the lies told, the evil acts that some people didn't hesitate to commit. I was frightened by the grandiose scheme that was about to surface.

Dolores's Sorrow

Over time Malcolm's mother Dolores came to me several times pleading for answers on whether I thought Malcolm's near-death illness in May 2009 could also have been from a plague infection.

I explained to her in layman's terms, "We have no blood samples saved from Malcolm in May to show that Malcolm had been infected with the same plague bacteria. Because he survived, there was no reason to keep his blood."

Dolores was referring to a near-death experience that Malcolm had suffered in May 2009 that also began with flu-like symptoms. In that twist of events, the doctors diagnosed a lacerated toe coincidental to the infection and treated his toe infection with antibiotics for eight weeks. Malcolm recalled the event in his home computer, noting how sick he was and not able to return to the laboratory for nearly two months. A FDA employee confidentially acknowledged to me that was the first time, in May 2009, that Malcolm was given the poison vaccine.

I regret that I did not give Dolores a reasonable answer regarding Malcolm's illness in May 2009. Dolores had the saddest look on her face. For whatever reasons, she wanted an answer for her son's death. Two years after Malcolm's death, Dolores passed away, not knowing what had happened to her beloved eldest child.

10

BIOTERRORISM PROTOCOL

As I continued to explore the facts and notions contained in the FOIA documents, I soon realized that the weaponry was not in the bacteria of KIM D27 itself but in the hands of scientists who target delivered the culprit to human blood. It was weaponry to induce septicemic plague and steal the lives of the infected.

While federal investigators descended upon the University of Chicago campus for the official investigation, I pursued answers on my own, my primary question being, how was it done? That is, how did a seemingly "harmless" pathogen, as Schneewind and Alexander would call it, reach Malcolm's blood, where it besieged him and took his life? Kathy Ritger had talked about a "two-day infection." What did that mean? I decided I had to answer two questions:

- How was KIM D27 directed to Malcolm's blood?
- What was in the witch's brew delivered in the "two-day infection" protocol?

Target Delivery to Induce Septicemic Death

I read an email sent from Dr. Paul Mead, a member of the investigation team, to Kathy Ritger prior to his arrival in Chicago. Dr. Mead was an M.D., MPH, and chief of epidemiology and surveillance at the National Center, CDC, Fort Collins, Colorado. On Saturday of September 19, 2009, at 3:39 p.m., just six days after Malcolm died, Dr. Mead had examined Malcolm's lung tissue slides and determined that it was not a pneumonic plague that had killed him (see Figure 10A).

> "Based on the clinical picture and my rudimentary interpretation of the lung tissue slides (albeit of a smaller area of the lung), I suspect that this was not the pneumonic exposure. Therefore, it would be most valuable if the pathologist could pay particular attention to identifying other potential portals of entry – e.g., needle sticks, cuts or sores on the hands, or possibly inoculation through the eye or through the GI tract." (Emphasis added, Figure 10A)

Dr. Mead went on to suggest alternative portals of entry (targeted delivery) consistent with Malcolm's septicemic death, namely "needle sticks, cuts or sores on the hands, or possibly inoculation through the eye or through the GI tract".

However, on the same day, Dr. Bryant, who performed an autopsy for the family, had ruled out these other portals of entries except the GI tract (gastrointestinal tract) delivery.

From: Mead, Paul (CDC/CCID/NCZVED) [mailto:pfm0@cdc.gov]
Sent: Sat 9/19/2009 3:39 PM
To: Ritger, Kathleen
Cc: Schriefer, Martin (CDC/CCID/NCZVED); Petersen, Jeannine (CDC/CCID/NCZVED)
Subject: RE: pathology

Kathy,

Based on the clinical picture and my rudimentary interpretation of the lung tissue slides (albeit of a small area of the lung), I suspect that this was not a pneumonic exposure. Therefore, it would be most valuable if the pathogologist could pay particular attention to identifying other potential portals of entry – e.g., needle sticks, cuts or sores on the hands, or possibly inoculation through the eye or through the GI tract. In addition to looking for skin lesions, this would involve looking for areas of regional lymphadenopathy for these respective sites. What to sample depends on what he sees during the gross examination, but in particular looking for and sampling enlarged regional lymph nodes. Beyond that, liver, spleen and lung are typically where we look for the organisms.

Does that help?

Paul

Figure 10A: email from Dr. Paul Mead to Kathy Ritger on his examination of Malcolm's lung tissue slides, pneumonic plague had been ruled out as cause of death.

So how was the pathogen delivered to Malcolm's blood? I labored over various methods that would mimic the delivery in the animal model study. Based on the information in Dr. Mead's email and the autopsy by Dr. Bryant, the answer seemed clear: the GI tract, the oral ingestion and the gastrointestinal route of delivery, could bring KIM D27 across the intestinal epithelium to the bloodstream.

Oral Vaccine Delivery

In his article "The Pathogenesis, Clinical Implications, and Treatment of Intestinal Hyperpermeability" (*Alternative Medicine Review* 2[5], 1997: 330–345), Allan Miller, N.D., writes:

> The gastrointestinal tract is unique in that it is an infolding of our outer skin. It is a continuous tube from the mouth to the anus, which allows the passage of nutrients into and through the body. The human gastrointestinal tract is in contact with food and antigens from the outside world, yet it is also intimately in contact with the interior of the body and the bloodstream.

I researched another article on "Oral Vaccine Delivery," by G. J. Russell-Jones, in the *Journal of Controlled Release* 65 (2000): 49–54, which explained the process in layman's terms. Russell-Jones noted that the attenuated pathogenic bacteria that gain access to human body via oral route specifically target the intestinal M Cells, a process known to transport pathogens to blood. Many researchers have therefore concentrated on identifying molecules, or pathogens, that target to the intestinal M cells.

The cartoon I have drawn below showed how pathogenic bacteria gain access to human blood (Figure 10B). The space between the intestinal epithelium marks the inner tube of the intestinal tract where food and nutrients pass through the gastric space. Red devils represent pathogens of attenuated bacteria, and the arrows track the movement of the pathogen *in transit* from intestinal tract to the bloodstream).

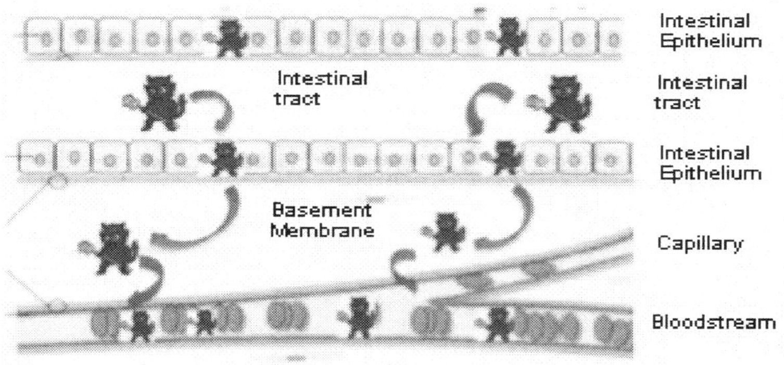

Figure 10B: Schematic diagram of pathogen to invade into capillary from intestinal tract

Immunologists who tracked the movement of the pathogen noted that pathogens targeted M cells located in the intestinal epithelium. The damaged M cells created lesions that could be viewed microscopically. The lesions created allowed pathogens to move across the epithelium, through the basement membrane, and enter the bloodstream capillaries that lie just underneath. Once in the blood, the pathogen multiplied to a full-blown plague and death ensued. It was like a "lethal injection" when the plague agent reached blood (see Figure 13H, NIH-RAC, June 16, 2010).

The window of opportunity to rescue the infected individual through antibiotic treatment is the first twenty-four hours after the initial symptoms appear. It was on September 4 that Malcolm first experienced his symptoms. That day, he attended the university's IBC monthly meeting at 1:45 p.m., as verified by his written notes on the meeting minutes later found in his office by CDC. In late afternoon, he began to feel sick. Schneewind sent Malcolm home to rest with no antibiotics, according to Kathy Ritger, who questioned the laboratory personnel. Because Malcolm received no antibiotic treatment within those first twenty-four hours when his symptoms first appeared, his death was inevitable from that day forward.

Bioterrorism Warfare Development Part I: Two-Day Infection Mix Delivered Intragastrically to human subjects

The delivery mechanism that culminated in Malcolm's death was oral ingestion and intragastric delivery through the digestive system, which allowed KIM D27 bacteria to gain access to blood. I searched the papers I had acquired, looking for a protocol that involved this delivery mechanism.

In 2006, Alexander Chervonsky, M.D., Ph.D., a faculty member in the Department of Pathology at the University of Chicago, developed a protocol that became part of the bioterrorism project. Chervonsky noted in his protocol that the administration of treated (boiled in water) and untreated *E. coli* to mice by gavage (that is, the substance was introduced to the stomach via a tube inserted down the throat) created lesions in the intestinal wall through villous M cells that allowed pathogens to go through and enter the blood in a two-day infection protocol.

This study was embodied in the protocol submission to NIH on July 25, 2007 (627-02). Schneewind noted in his document submission, "Oral administration of these reagents (0.5 ml/mouse) is the only way to put them in direct contact with intestinal flora and epithelium." His idea was in line with the oral vaccine delivery in Russell-Jones's article to direct pathogens to the human bloodstream via intestinal M cells.

As a basis for the pathogen delivery, Schneewind and Chervonsky successfully developed a protocol to bring pathogens to the blood via villous M cells located in intestinal epithelium. In his protocol, Schneewind noted the following (Figure 10C, Appendix 1, p9-10, emphasis added):

> "To study the regulation of villous M cell development, we will use oral gavage with antibiotics (streptomycin) or heat-treated non-pathogenic bacteria (*E. coli* DH5-alpha) strain. Chervonsky lab has found that *E. coli* DH5-alpha induce intestinal M cells when orally administered to mice after exposure to water and one minute of boiling. Oral administration of these reagents (0.5ml/mouse) is the only way to put them in direct contact with intestinal flora and epithelium. Preliminary experiments in Chervonsky lab [IACUP 7361] found a single dose of 0.5g/kg of streptomycin to be non-toxic for 25g BALB/c mice. To examine the role of induction of additional M cells by stressed commensal flora (*E. coli* lysates in water) in mucosal immunity, we will challenge mice intra-gastrically with Y. enterocolitica via oral gavage. Mice previously gavaged (at 24–36 hrs time point) with 20 ODU/ml *E. coli* in water, 20mg/mouse of Streptomycin, or PBS control will be infected with up to 10^9 CFU of Y. enterocolitica intragastrically by oral gavage (0.5ml).
>
> Mice will be sacrificed at 24, 48, 72 and 96 hrs and at later points if needed (day 5, 7, 9, 14) after infection. The number of bacteria in the liver, spleen and small intestine will be determined. Mice will be observed immediately after oral gavage to be sure that they do not have liquid in their lungs." (Emphasis added, Figure 10C, Appendix 1)

oral gavage. Mice previously gavaged (at 24-36 hrs time point) with 20 ODU/ml E. coli in-water, 20mg/mouse of Streptomycin, or PBS control will be infected with up to 10e9 CFU of Y. enterocolitica intragastrically by oral gavage (0.5 ml). Mice will be sacrificed at 24, 48, 72, and 96 hrs and at later time points if needed (day 5, 7, 9, 14) after infections. The number of bacteria in the liver, spleen, and small intestine will be determined. Mice will be observed immediately after oral gavage to be sure that they do not have liquid in their lungs. Mice will be monitored daily after

To study the regulation of villous M cell development, we will use oral gavage with antibiotics (streptomycin) or heat-treated non-pathogenic bacteria (E. coli DH5alpha strain). Chervonsky lab has found that E. coli DH5aplpha induce intestinal M cells when orally administered to mice after exposure to water and one minute of boiling. Oral administration of these reagents (0.5 ml/mouse) is the only way to put them in direct contact with intestinal flora and epithelium. Preliminary experiments in Chervonsky lab [IACUP 73641] found a single dose of 0.5g/kg of streptomycin to be non-toxic for 25g BALB/c mice. To examine the role of induction of additional M cells by stressed commensal flora (E. coli lysates in water) in mucosal immunity, we will challenge mice intra-gastrically with Y. enterocolitica via oral gavage. Mice previously gavaged (at 24-36 hrs time point) with 20 ODU/ml E. coli in-water, 20mg/mouse of Streptomycin, or PBS control will be infected with up to 10e9 CFU of Y. enterocolitica intragastrically by oral gavage (0.5 ml). Mice will be sacrificed at 24, 48, 72, and 96 hrs and at later time points if needed (day 5, 7, 9, 14) after infections. The number of bacteria in the liver, spleen, and small intestine will be determined. Mice will be observed immediately after oral gavage to be sure that they do not have liquid in their lungs. Mice will be monitored daily after

Figure 10C: Oral gavage protocol – two-day infection regime for Bioterrorism, Part I

A Two-Day Infection Regime

In other words, a poison mix, a witch's brew, is introduced during a two-day infection process like the one Kathy Ritger had discussed with Brooke and me in Malcolm's case.

I. *E. coli* lysates: <u>Oral ingestion</u> of 20 OD units/ml of *E. coli* DH5-Alpha strain mixed in water with 20mg/mouse of streptomycin

II. 24-36 hours later, the infection mix was delivered orally.

III. Infection mix: Oral ingestion of 10^9 CFU of *Yersinia enterocolitica*, in PBS buffer

Table 1: Infection mix delivered by oral ingestion through gastrointestinal tract.

Bioterrorism Warfare Development Part II: Human Trials with Test Pathogen, Yersinia enterocolitica

On July 25, 2007, Schneewind and Chervonsky devised a protocol for human trial using heat-treated *E. coli* DH5-alpha lysate in water mixed with streptomycin in oral and intragastric delivery to human subjects. Twenty-four hours later, *Y. enterocolitica* as test pathogen was delivered orally to the human subjects. As the protocol noted, "Human infections with *Y. enterocolitica* are not frequent," which indicated its relatively benign and safe nature. Both Schneewind and Chervonsky are medical doctors who had access to antibiotics when needed in case of an infection.

And the dosage noted in the protocol was 10^2–10^9 pfu (plague-forming units) per kg of body weight of animals or humans that was contained in a volume of 0.5ml for oral/intragastric delivery.

> "What are the symptoms of an infection [via oral delivery]?" asked the protocol administrator.
>
> "Yersinia enterocolitica can infect the intestines of humans by oral contamination and can cause Intestinal lymph adenitis as well as damamge [damage] the intestinal epithelium. These lesions cause diahrrhea [diarrhea] and/or vomiting as well as fever." (Emphasis added, Figure 10D, Appendix 1).

What are the symptoms of an infection?

Yersinia enterocolitica can infect the intestines of humans by oral contamination and can cause intestinal lymph adenitis as well as damamge the intestinal epithelium. These lesions cause dialrrhea and/or vomiting as well as fever. Human infections with Y. enterocolitica are not frequent;mostly infants and immunocompromised

Figure 10D: Symptoms of pathogen delivery via intragastric route (Bioterrorism part II

It is clearly indicated that the oral delivery protocol with test pathogen would cause diarrhea, vomiting, and fever—the same symptoms suffered by Malcolm, as would be noted in his autopsy report. Also noted in the autopsy report would be "damage and lesions to the intestinal epithelium" (see Chapter 11).

What followed in this protocol sent shivering waves of shock through me as I continued to read the protocol recorded on July 25, 2007 (Figure 10E, Appendix 1 & 2).

> "How can staff be exposed to the agent? (i.e., ingestion
> or other percutaneous contact ....)?"
>
> "Ingestion and direct injection into the bloodstream"

It was the beginning of a human trial to direct pathogens to blood to induce septicemic death. It was biowarfare at work. The mystery of a plague infection that resulted in septicemic death had finally been revealed. For pathogens to enter the bloodstream, Schneewind cited two options (Figure 10E):

- Ingestion via intragastric route
- Direct injection into the bloodstream.

Undoubtedly, both delivery mechanisms would bring death to the victims. Oral ingestion had the advantage of directing an internalized infection without external appearances. Direct injection into the bloodstream, though most direct and a lethal injection, would leave visible scab wounds at the injection site thereby alerting health professionals to identify the infection.

In the event of exposure, the PI/supervisor (Schneewind), UCOM, and ARC were to be immediately notified (Figure 10E, Appendix 1). I recalled the email that Malcolm sent to Schneewind and Blaylock on September 10 when he was near death. It went unanswered.

*****

"Will staff be monitored for infections?"

"No, Surveillance is not appropriate for this agent" (Figure 10F)

*****

"Is pre-exposure vaccination available?"

"No," was the answer. But noted,

"Please contact UCOM to make arrangements for vaccination." (Figure 10G, arrow, emphasis added).

*****

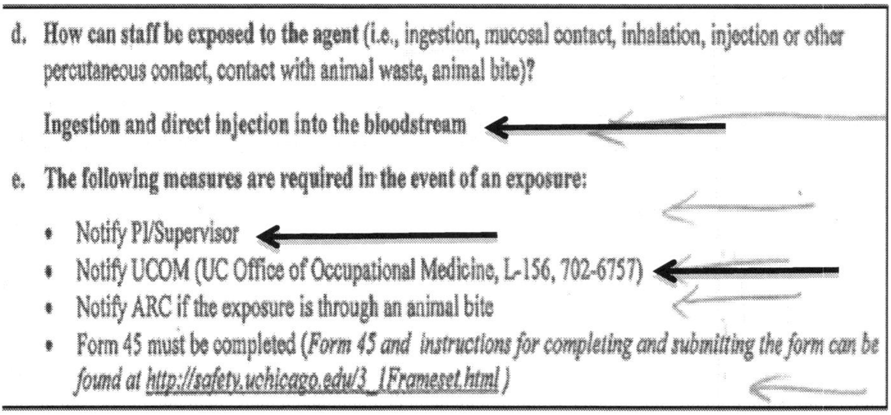

d. **How can staff be exposed to the agent** (i.e., ingestion, mucosal contact, inhalation, injection or other percutaneous contact, contact with animal waste, animal bite)?

Ingestion and direct injection into the bloodstream

e. **The following measures are required in the event of an exposure:**

- Notify PI/Supervisor
- Notify UCOM (UC Office of Occupational Medicine, L-156, 702-6757)
- Notify ARC if the exposure is through an animal bite
- Form 45 must be completed (*Form 45 and instructions for completing and submitting the form can be found at* http://safety.uchicago.edu/3_1Frameset.html)

Figure 10E: Staff exposed to pathogen by ingestion and direct injection into the bloodstream in a human trial, Bioterrorism, part II

8. **Will staff be monitored for infections?**

☒ No: Surveillance is not appropriate for this agent

☐ Yes: **Please explain.**

Figure 10F: Staff involved in the human trial of plague Infection will not be monitored for infection, Bioterrorism, Part II

As I continued to read the document, the vaccination took place at UCOM as a footnote in the bottom of Section 6 of the document.

6. **Is pre-exposure vaccination available?**

☒ No

☐ Yes*: **Please specify.**

*Please contact UCOM to make arrangements for vaccination.

Figure 10G: Staff and researchers in the human trial are to contact UCOM to make arrangement for vaccination.

Finally, the plague vaccination program had emerged at the University of Chicago. The protocol requested human subjects to make arrangement for vaccination at the University of Chicago Occupational Medicine (UCOM), headed by Dr. Geoffrey Korn (Figure 10G, Appendix 1).

What appeared before me was a protocol constructed specifically for the delivery of biological weapons to human subjects. It was a mechanism to kill, a silent assassin, not just of lab animals, but of human subjects, such as researchers forced to participate in a study.

Eight members of the staff, Schneewind and Chervonsky included, signed and dated the protocol for human trials on November 8, 2007 (Figure 10H, Appendix 1). I noticed that Bill Blaylock and Schneewind were both in this group, the men Malcolm notified of his illness on September 4 and the 10 just days before his death.

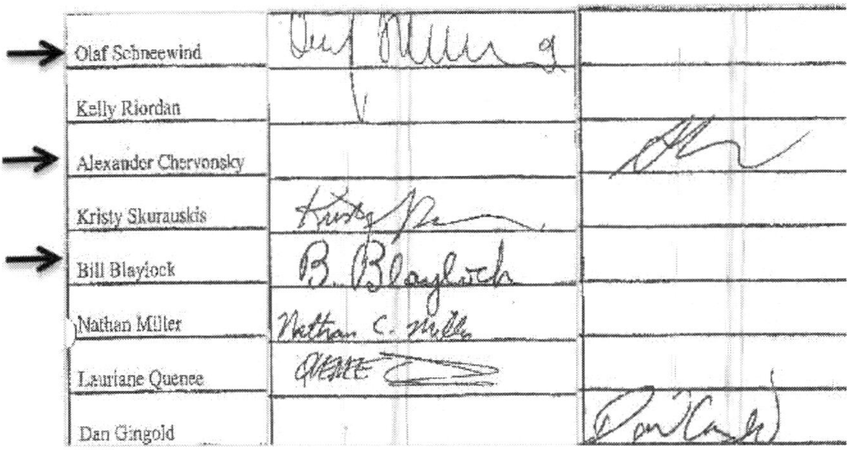
Olaf Schneewind
Kelly Riordan
Alexander Chervonsky
Kristy Skurauskis
Bill Blaylock
Nathan Miller
Lauriane Quenee
Dan Gingold

Figure 10H: Human trial candidates and their signature in Bioterrorism, Part II

Bioterrorism Warfare Development: Part III: Human Trials using KIM D27-Cm^R

After the successful launching of a protocol to cause septicemic blood infection, Schneewind was in good position to initiate human trials with the deadly pathogen *Yersinia pestis*, KIM D27-Cm^R.

"Plague at University of Chicago"

On October 7, 2008, Schneewind submitted a protocol submission to NIH titled, "Plague at University of Chicago" (Appendix 2) Schneewind requested the use of *Y. pestis, KIM D27 (RG2/BSL2)* in his human trials as noted in his Specific Aim 5, on the protocol submission, "Vaccine candidates for Y. pestis." (Figure 10I). The stage was set for a drama that would soon bring Malcolm down.

Specific Aim 4. Immune response to Y. pestis infection.
Specific Aim 5. Vaccine candidates for Y. pestis.

Figure 10I: Bioterrorism Protocol specified vaccine candidates for Y. pestis

- **Bioterrorism Agent - <u>KIM D27 (RG2/BSL2).</u>**

And, Schneewind noted in his opening statement.

> "*Y. pestis* remains a major threat for bioterrorism and biological warfare. Because Americans are not immunized against plague, we rely on basic research for the development of a vaccine, immunotherapies and therapeutics that can be administered to people that have been exposed to plague. At present, there is no vaccine available in the United States."
>
> "<u>Initial screening will be performed by using the attenuated (non-virulent) *Y. pestis* KIM D27</u> (<u>RG2</u>/<u>BSL2</u>)" (Figure 10J, Appendix 2, emphasis added).

Yersinia pestis remains a major threat for bioterrorism and biological warfare. Because Americans are not immunized against plague, we rely on basic research for the development of a vaccine, immunotherapies and therapeutics that can be administered to people that have been exposed to plague. At present, there is no vaccine available in the United States.

Initial screening will be performed by using the attenuated (non virulent) *Y. pestis* KIM D27 (RG2/BSL2).

KIM D27 (RG2/BSL2) = KIM D27-Cm^R

Figure 10J: Initial screening of bioterrorism human trial using KIM D27-Cm^R was noted.

In his bioterrorism protocol on October 7, 2008, Schneewind characterized his genetic manipulation of *Y. pestis.* The question that set me madly in pursuit in the last few months had just revealed itself plainly and explicitly before my eyes (Figure 10J, Appendix 2).

RG (risk group) and *BSL* (biosafety level) are terms used to assess the danger of a pathogen.

• Bioterrorism Agent - KIM D27-Cm^R

The transfer of chloramphenicol drug resistance onto the genome of KIM D27 is a common technique used in standard molecular biology. It enabled clonal selection of culture when grown in medium containing chloramphenicol.

KIM D27 (RG2) strain that specifies resistance to chloramphenicol was the strain used in the human trial according to the book of bioterrorism (Figure 10K, 10L).

In the "Plague at University of Chicago, Schneewind noted.

> "Please note that all *Y. pestis* RG3, antibiotic resistance markers used for genetic manipulation will be Kanamycin or Ampicillin. Chloramphenicol will not be used in RG3 but will be used in RG2 *Yersinia* (attenuated strains)." (Figure 10K, Appendix 2, Emphasis added)
>
> "Chloramphenicol for KIM (RG2) strains." (Figure10L, Appendix 2, emphasis added)

• **Genetic manipulation of *Y. pestis***

Please note that for all *Y. pestis* RG3, antibiotic resistance markers used for genetic manipulation will be Kanamycin or Ampicillin. Chloramphenicol will not be used in RG3 but will be used in RG2 *Yersinia* (attenuated strains).

Figure 10K: Chloramphenicol will not be used in RG3 but will be used in RG2 Yersinia attenuated strains

Will this research involve the deliberate transfer of a drug resistance trait to a pathogenic organism (pathogenic to humans, animals or plants) for which no alternative drugs are available?

☐ No

☒ Yes, please list the antibiotic resistance markers (be sure to include this information in the appropriate Agent Profile Form): Kanamycin/Ampicillin/Chloramphenicol for KIM (RG2) strains
Kanamycin/Ampicillin for RG3 strains

Figure 10L: Schneewind designated a KIM D27 derivative, KIM D27-Cm^R, (RG2/BSL2) that is resistant to chloramphenicol to be used in the Bioterrorism human trial, part III.

- **Isolate Recovered From Malcolm's Blood (UC91309) Shares Total Sequence Identity With Bioterrorism Agent, KIM D27-Cm^R**

As Dr. Shriefer of the CDC investigation team noted:

> "The strain characterization performed by Dr. Schriefer has confirmed that initial characterizations performed in the Schneewind lab on 9/18/09. Addition subsequent experiments performed in the Schneewind and Schriefer labs have further characterized this clinical isolate as resistant to chloramphenicol, but sensitive to all other antibiotics tested." (Emphasis added, Figure 10M)

the strain characterization performed by Dr. Schriefer has confirmed that initial characterizations perfomed in the Schneewind lab on 9/18/09. Additional subsequent experiments performed in the Schneewind and Schrieifer labs have further characterized this clinical isolate as resistant to chloramphenicol, but sensitive to all other antibiotics tested via Kirby-Bauer and/or Microbroth dilution assays (kanamycin, streptomycine, gentamycin, tetracycline, ciprofloxacin, and ampicillin ,doxycycline and levofloxacin).

Figure 10M: Isolate recovered from Malcolm's pre-mortem blood was chloramphenicol resistant

In Incidence Report to NIH-OBA on November 23, 2010, Schneewind and Kanabrocki noted that sequencing determined by Northern Arizona University on the pre-mortem isolate also shown to contain chloramphenicol cassette marker inserted into the KIM D27 genome that specified chloramphenicol resistance. The insertion site showed that it was the same strain constructed by Malcolm as indicated in his lab notebook. (Figure 10N).

- "Genome sequencing of UC91309 also identified the insertion site of a cassette encoding Chloramphenicol Acetyl Transferase (cat) of Y. pestis."
- "UC91309 was derived from Y. pestis, KIM D27, via insertion of this cat cassette"
- "Laboratory notebooks of the Deceased included a description of the oligonucleotide primers synthesized to facilitate insertion of the cat cassette." (Figure 10N, Emphasis added).

> The Deceased was engaged in research involving the attenuated, iron acquisition defective strain *Yersinia pestis* KIM D27, which harbors a *pgm* deletion. Genomes of both *Yersinia pestis* strains, KIM D27 and UC91309, were sequenced. Genome sequencing of UC91309 and KIM D27 confirmed the absence of the *pgm* locus and the High Pathogenicity Island, a genomic region bearing the locus for Yersiniabactin, a siderophore known to be an iron scavenger of transferrin. Genome sequencing of UC91309 also identified the insertion site of a cassette encoding Chloramphenicol Acetyl Transferase (*cat*). *Y.p.* UC01309 was derived from *Y.p.* KIM D27 via insertion of this *cat* cassette. Laboratory notebooks of the Deceased included a description of the oligonucleotide primers synthesized to facilitate insertion of the *cat* cassette.

Figure10N: Genomic sequencing of UC91309 showed the construct had the chloramphenicol cat cassettes insertion

Bioterrorism Warfare Development: Part IV: Human Trial using KIM D27-Cm^R

A KIM D27 derivative that specified chloramphenicol drug resistance to the bacteria KIM D27 (RG2/BSL2) was, indeed, constructed by Malcolm (Figure 10N). The chloramphenicol tagged KIM D27 was later designated by Schneewind to be used in the Bioterrorism human trial (RG2) shown in his protocol submission to NIH on October 7, 2008. Explicitly, strain isolate from Malcolm's pre-mortem blood that took his life shared total sequence identity to the designated Bioterrorism Warfare Agent, KIM D27-Cm^R (Figure 10K, 10L, 10M, 10N).

The fact that KIM D27-Cm^R was the strain designated for use in the human trial on October 7, 2008; and the isolate recovered from Malcolm had shared total sequence identity leads to the inevitable conclusion that Malcolm was killed by the bioterrorism warfare protocol and the bioterror agents developed in Schneewind's laboratory.

But the fact made the case even more bizarre. Malcolm was killed by the very strain that he himself constructed in the acquisition of chloramphenicol cat cassettes that specified chloramphenicol resistance to KIM D27. Indeed, it was the bioterrorism weaponry that was used to take Malcolm's life by Schneewind.

This horrific detail threw me into an unimaginable feeling of defeat, crumbling me to the floor like a deflated animal at the masquerade party. I was stumped upon by those who wore masks. There were laughers and jeers, whispers and trumpets, all chanted triumph in the House of Bioterrorism.

A retaliation effort so dreadful and despicable that it could only be conceived of by the lowest human criminals on earth, but had actually been carried out by those who wore academic hoods and gowns in University endowed Professorship and who spoke of honesty and integrity to academic students. And they had carried it out against Malcolm, a man who had done nothing wrong in his entire academic life, who had spent his career pursuing science in its purest sense. Wasn't Malcolm's death a reflection of the story and death of Howard Taylor Ricketts who injected himself with the virulent material that he purified from his research (T. W. Goodspeed, "Howard Taylor Ricketts," *The University Record*, April 1922: 105–106).

My tears had gone dry by this time, replaced by a mounting anger. I could hear Malcolm's cry, "Vaccine they gave me, Poison! ... Poison! ..."

I continued to read the FOIA documents that told the story.

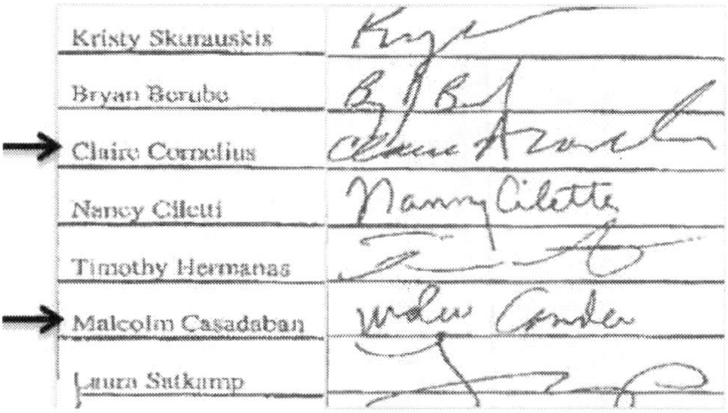
Kristy Skurauskis
Bryan Berube
Claire Cornelius
Nancy Cilletti
Timothy Hermanas
Malcolm Casadaban
Laura Sutkamp

Figure 10P: Human candidates certified to receive mandatory vaccination in Bioterrorism part IV.

Seven people were acknowledged as participants in the human trial dated on August 21, 2008; they included Malcolm and Claire Cornelius (see arrows), the army major from DOD. They were "certified" new members to Schneewind's laboratory (Figure 10P) for one reason only—to engage in a deadly human trial.

Bioterrorism Warfare Development
Part V: Malcolm Besieged by U.S. Bioterrorism

When the curtain fell, Malcolm was on a stretcher wrapped in a white blanket in *exitus lethalis*. I took a deep breath and sighed. Images of Malcolm in the morgue continued to flash before me as if I had only seen him there yesterday.

1. Malcolm became a certified new employee to Schneewind's GLRCE, a National Lab, engaging in bioterrorism research on August 21, 2008. This was a demotion against his contracted tenure at the University of Chicago held since 1987. I recalled the promise that Keith Moffat made to him on June 26, 2009, that nothing of this sort would happen to him. The certification compromised his standing at the university, obligating him to take part in the deadly human trials that were part of the bioterrorism protocol on August 21, 2008 (Part IV). And he died on September 13, 2009.

2. KIM D27 (RG2/BSL2) that specified resistance to chloramphenicol drug was the designated bioterrorism warfare agent to be used in the human trial according to the protocol submission on October 7, 2008 (Figure 10K, 10L). The strain isolate found in Malcolm's pre-mortem blood shared sequence identity with the one designated in the bioterrorism protocol for human trials (Figures 10M, 10N).

3. Bioterrorism protocol dated July 25, 2007, (Part I) specified a mechanistic "two-day infection" regime in the delivery of pathogens to human subjects. The poison mix that contained KIM D27-Cm^R was actually delivered to Malcolm on the second day of the two-day infection regime, specifically, on September 1, 2009, according to Kathy Ritger of the federal investigator team.

4. Malcolm suffered diarrhea, vomiting, and fever in the course of his infection predictable from an oral and intragastric delivery of Y. enterocolitica in the human trial noted in bioterrorism protocol (Part II).

5. Lesions and gastrointestinal damages noted, as "intestinal mucosal prolapse with bacterial foci" in the autopsy report was consistent with symptoms when *Y. enterocolitica* was used as test pathogen as in the bioterrorism protocol (Part II). Intestinal lesions and intragastric damages reflected injuries by pathogens moving across intestinal epithelium to enter into the bloodstream.

6. Code of death, "untreated septicemic plague are fatal" and "*exitus lethalis*" cited in bioterrorism protocol on October 8, 2008, and Biosafety Manual marked the deadliness of the pathogen that was used to inoculate Malcolm. It was the basic framework set to besiege Malcolm in the house of U.S. bioterrorism.

7. Mechanistic delivery of pathogens to human blood by way of intragastric delivery (gastrointestinal delivery) induces septicemic death in bioterrorism protocol (Part I). It was also the cause of death according to Malcolm's autopsy report. It was the culmination of a deadly pathogen through target delivery that won

Schneewind and Chervonsky the award in the biological warfare development.

8. Malcolm was not treated for or rescued from his fatal infection on September 4 in Schneewind's laboratory and on September 10. Malcolm was the intended victim in "untreated pneumonic and septicemic are fatal" as in bioterrorism protocol submission to NIH on October 7, 2008.

9. Iron overload or genetic hemochromatosis was not the cause of death for Malcolm. This is because the blood environment of a normal healthy individual maintains iron level at 40-160 mcg/dl, (Figure 11G) which could support KIM D27 multiplication to the fullest extent. KIM D27 infection, thereby, is virulent in normal individuals irrespective of the extra iron brought by the hemochromatosis individuals. The actual cause of death was the mechanistic delivery of pathogens to Malcolm's blood that led to his septicemic infection and death.

10. The portal of entry, route of infection, and the intragastric delivery culminated in a septicemic infection that was fatal to humans, and Malcolm was the victim of the U.S. bioterrorism protocol.

11. The bioterrorism project acknowledged its funding support from the National Institute of Allergy and Infectious Disease (NIAID) with tracs ID #26020. It was the American taxpayers' money that paid for the bioterrorism project (Figure 10Q, Appendix 2).

B. Project is funded:

☐ Internally
☒ Externally
Awarded? ☒ Yes ☐ No
Grant Agency: NIAID
Tracs ID #: 26020

Figure 10Q: Bioterrorism Project funded by taxpayer's money through NIAID

I was devastated when the protocol revealed the nature of infection and path of human trials that killed Malcolm. It was the basis of bioterrorism weaponry disguised as vaccines that targeted Malcolm and took his life. It was a project sponsored by DOD in joint efforts with NIH, NIAID, FDA, and the National Lab at the University of Chicago that ran hundreds of millions of U.S. taxpayer dollars to build such biological weaponry (Chapter 4).

Claire Cornelius was an Army major from DOD who stationed in Schneewind's laboratory to oversee the completion of the bioterrorism project.

That system had been used to strike down a brilliant, caring science professor, a man who taught biology and microbiology to students and helped draft the Biosafety Standard Protocol for all biological laboratories on major U.S. campuses. He spoke of his concerns about the unsafe nature of bioterrorism products. He had done nothing to warrant his horrendous fate.

Yet NIH and the federal administration continued to support U.S. bioterrorism programs in the killing of laboratory researchers.

> "Acquisition of this knowledge enables the development of products designed to prevent, diagnose, and treat diseases caused by agents of bioterrorism or newly emerging pathogens (NIAID, category A-C agents)" (Figure 10R).

These were the words from NIH Director, Francis Collins, and President Obama when the ARRA (American Recovery and

Reinvestment Award) award to Schneewind and the University of Chicago was announced just three days after Malcolm's death.

THE UNIVERSITY OF
CHICAGO

NIH AWARD FROM THE NATIONAL INSTITUTE OF ALLERGY AND INFECTIOUS DISEASES

Molecular Analyses and Interventions for Biodefense and Emerging Pathogens

Principal Investigator: Olaf Schneewind, MD, PhD, Professor and Chair, Department of Microbiology; Director, Great Lakes Center for Excellence
Start Date: September 17, 2009
Total Award Amount: $911,792

Public Health Relevance

The National Institute of Allergy and Infectious Diseases (NIAID) established the RCE Network, to gain appreciation of the molecular mechanisms whereby microbial pathogens cause human disease. Acquisition of this knowledge enables the development of products designed to prevent, diagnose, and treat diseases caused by agents of bioterrorism or newly emerging pathogens (NIAID Category A-C agents).

Project Description

Our nation's ability to detect, prevent and counter bioterrorism and emerging infectious diseases depends on technologies that are generated through biomedical research on disease-causing microbes and the human immune system's response to them. The National Institute of Allergy and Infectious Diseases (NIAID) established the RCE Network, i.e. the Research Centers of Excellence for Biodefense and Emerging Infectious Diseases, to gain appreciation of the molecular mechanisms whereby microbial pathogens cause human disease. Acquisition of this knowledge enables the development of products designed to prevent, diagnose, and treat diseases caused by agents of bioterrorism or newly emerging pathogens (NIAID Category A-C agents).

This award is funded under the American Recovery and Reinvestment Act of 2009, NIH Award number: 3U54AI057153-06S1

Figure 10R: ARRA award from President Obama and NIH to the University of Chicago three days after Malcolm's death.

11

THE AUTOPSY REPORT

In December 2009, the long-awaited autopsy report finally came to us, forwarded by Kathy Ritger. It was a cold day just before Christmas, the first the family would celebrate without Malcolm. His loss cast a dark shadow, his unexplained death haunting us all. We counted the days, not to Christmas, but from Malcolm's death.

My hands trembled as I sat to read the autopsy report along with the medical history of Malcolm's last hours in the emergency room. Together they painted a gruesome picture of how Malcolm fared in his final hours. Reading the documents was almost more than I could bear, but I pushed forward. I had to find the truth.

Key points from the Autopsy Report (Appendix 2) and Medical History of his Emergency Room were summarized below.

I. Septicemic infection by KIM D27-Cm^R

Malcolm suffered a septicemic blood infection that led to his death. The culprit was a KIM D27 derivative that conferred chloramphenicol resistance to the bacteria isolated from his pre-mortem blood. The isolate bore identical genomic sequence with the KIM D27-Cm^R, the designated strain for use in the human trial in the bioterrorism protocol (Part III) on October 7, 2008.

II. Portal of Entry: The Gastrointestinal Tract

In an effort to determine the cause of death, the autopsy team rigorously sought the "portal of entry and route of infection" for Malcolm's death. The autopsy report in Appendix 5 noted:

> " - Area of mucosal prolapse of the sigmoid colon. This area is associated with a small area of possible mucosal damage or ulceration and with focal bacterial organisms (Gram positive on Gram stain)" (Figure 11A)

- Area of mucosal prolapse of the sigmoid colon. This area is associated with
a
small area of possible mucosal damage or ulceration and with focal bacterial
organisms (Gram positive on Gram stain).

Figure 11A: Autopsy Report on mucosal damage or ulceration with focal bacterial organisms

The mucosal damage or ulceration noted in the autopsy report was consistent with the intragastric delivery noted in bioterrorism protocol (Part II) of July 25, 2007 (see Figure 10C, 10D).

The autopsy report further concluded from all tests performed that the portal of entry through the gastrointestinal tract might be the cause of death in Malcolm (Figure 11B).

> "Based on these findings, the overall history and possibly the presence of focally ulcerated mucosal prolapse in the colon, the portal of entry through the gastrointestinal tract may have to be considered." (Emphasis added, Figure 11B)

Yersinia pestis (KIM D27), after extensive laboratory testing. The portal of entry or source of infection by this organism is unclear. There were no skin lesions and there was no morphologic evidence of pneumonia. No larger ulcerated lesions were found in the oropharynx or gastrointestinal tract. Based on theses findings, the overall history and possibly the presence of focally ulcerated mucosal prolapse in the colon, the portal of entry through the gastrointestinal tract may have to be considered.

Figure 11B: Portal of entry defined as cause of death in Malcolm's Autopsy Report

III. Intestinal Epithelium Damage—Diarrhea, Vomiting, and Fever

Oral ingestion and intragastric delivery of the pathogen incurred damages to the intestine and caused a variety of symptoms, including diarrhea, vomiting, and fever—those noted in the protocol submission on July 25, 2007 (see Chapter 10, 10B, 10C and "Bioterrorism Warfare Development," Part II).

The emergency room professionals noted the same symptoms in Malcolm's medical history. In particular, they listed the diarrhea arising from damages to the intestinal epithelium and the fever and severe headaches and body aches that had tormented Malcolm for several days and continued into his final hours. All were derived from the bacterial toxins that infiltrated his brain, blood, and peripheral tissues (Figure 11C, 11D).

By the end of the day, Malcolm was suffering hyperlipidemia (too many lipids, or fats, in the blood), dry cough, fever, chills, leg edema, left lower extremity swelling, and abdominal swelling. On top of it all, he experienced shortness of breath, labored breathing that worsened with each passing hour in the ER. Later that evening, his body functions began to shut down by congestive heart failure, unspecified renal failure, severe headaches and body aches throughout until he succumbed to death at 7:08 p.m., on Sunday evening, September 13, 2009 (Figure 11D).

WATTS R.N., LA-RHONDA (REGISTERED NURSE) 09/13/2009 4:58 AM EMERGENCY MEDICINE

- Pt c/o sob bodyaches fever seen by ED Resident nad pt placed on monitor blood drawn and sent o2 100% non rebreather dexi level 122 rr 36 physcian aware CXR

PAIN ASSESSMENT					
Pain?	Yes -LW 09/13/09 0453	Yes -LW 09/13/09 0510	Yes -LW 09/13/09 0558	Yes -LW 09/13/09 0633	Yes -LW 09/13/09 0659
PAIN DESCRIPTION					
Location	Head -LW 09/13/09 0453	Head -LW 09/13/09 0510	Head -LW 09/13/09 0558	--	Head -LW 09/13/09 0659
Frequency	--	--	--	--	Continuous -LW 09/13/09 0659
Onset of Pain	--	Days -LW 09/13/09 0510	--	--	--

Row Name	09/13/09 1458
ABDOMINAL ASSESSMENT	
GI/GU/GYNE Assessment	Exceptions -AK 09/13/09 1459
GI	
GI Symptoms	Diarrhea -AK 09/13/09 1459

ASSESSMENT	
Skin Color	Cyanotic -AK 09/13/09 1459

Associated Diagnoses

SHORTNESS OF BREATH [786.05]
CONGESTIVE HEART FAILURE, UNSPECIFIED [428.0]
RENAL FAILURE [586M]

Figure 11C: Diarrhea and Cyanosis were noted in the ER before Malcolm died

History of Present Illness

HPI Comments: Ptn pmhx dm, hyperlipidemia with shortness of breath x 1 week worse tonight, + dry cough ptn with fevers, chills, and weakness over the same amount of time. Associated with pnd, orthopnea, leg edema, left lower extremity swelling, and abdominal swelling. He denies h/o chest pain, cad, no prior events

Difficulty Breathing

The history is provided by the patient. Associated symptoms include shortness of breath.

Figure 11D: Headaches, body aches and complete body shutdown before Malcolm succumbed to death

There were signs of cyanosis throughout his body earlier in the ER that worsened so that his body was deep blue by the time he expired. Cyanosis is a sign, in medical terms, for lack of oxygen in the blood, and the lack of oxygen led to shortness of breath, difficult breathing, and finally suffocation (Figure 11C).

These symptoms confused me. Difficult breathing and cyanosis are not symptoms of septicemic plague death. Difficulty breathing can be a symptom of pneumonic plague, but Dr. Mead as well as the doctor performing the autopsy had ruled out pneumonic plague after examining Malcolm's lung tissue slides (see Figures 10A). There had to be another mechanism at work, something else that contributed to Malcolm's death. What was it?

IV. Streptococcus bacteria and Nutritional Variant Streptococcus (NVS)

As Schneewind and others had told us, and the autopsy report confirmed, Malcolm's pre-mortem blood had more than one bacterial species present. One in particular was the Nutritionally Variant Streptococci (NVS), a derivative and a spinoff satellite colony derived from the helper bacteria *Streptococcus*.

Seeing this, I had two immediate questions:

- Was the hemolytic activity of *Streptococcus* responsible for Malcolm's death?
- How was *Streptococcus* bacteria introduced to Malcolm's blood when blood is not its natural reservoir?

Further research answered the first question.

Streptococcus bacteria are classified into three major groups based on their hemolytic activities exhibited on the red blood cells. The alpha group (e.g., the *S. viridans*, right in Figure 11E) causes oxidization of iron in hemoglobin molecules in the red blood cells, giving rise to greenish color on blood agar by the iron released from the red blood cells to the medium. Beta hemolytic species (e.g., *S. pyogenes*, left in Figure 11E) causes complete rupture of the red blood cells, thereby creating a clear halo indicating dead red blood cells in the center of colony. For both alpha and beta hemolytic activities, iron would be released into blood. Gamma hemolytic species does not cause rupture of the red blood cells and shows no iron release to the medium.

http://en.wiki.pedia..org/wiki/Streptococcus

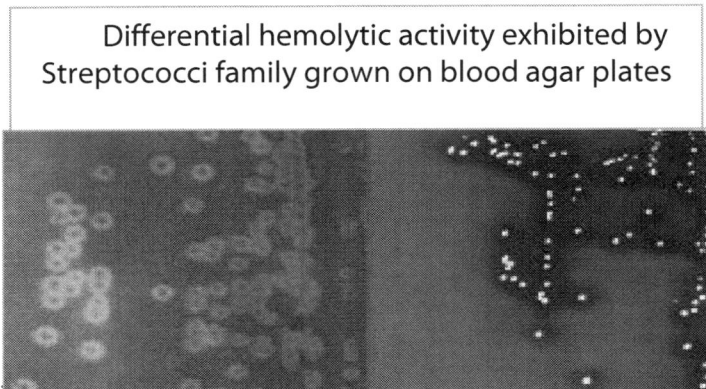

Differential hemolytic activity exhibited by Streptococci family grown on blood agar plates

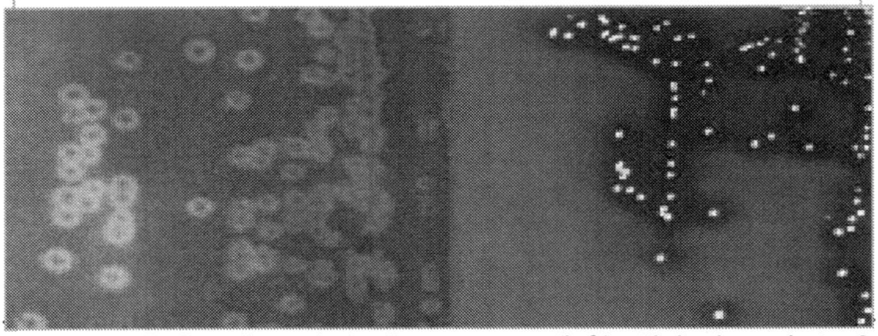

Figure 11E: Beta-hemolytic S. pyogenes (left), Alpha-hemolytic S. viridans (right)

V. Hemolytic Activity of Streptococcus bacteria

Significantly many species of the *Streptococcus* genus of bacteria can cause hemolysis, which involves the dissolution and rupture of the red blood cells. When the red blood cells rupture, they release hemoglobin into the serum. This disrupts the ability of the red blood cells to carry oxygen through the body. As the hemolytic activity continues, there are fewer and fewer red blood cells able to circulate oxygen. With increasing oxygen depletion, difficulty breathing develops, and the sufferer's skin gradually turns a dusky blue (cyanosis). The person slowly suffocates and dies.

VI. Hemolytic Activity Contributes to Malcolm's Death

In order to determine and differentiate which Streptococcus species were present in Malcolm's blood, I looked at his medical history.

The blood work presented in Figure 11F showed a 28 percent reduction of red blood cells (RBC) at 3.74; the normal range is 4.47–5.91 M/ul (million per microliter). This reduction could explain the shortness of breath led by the hemolytic activity associated with the *Streptococcus* bacteria. White blood cells (WBC), by contract, were elevated to a high level, suggesting his immune function was fully charged to combat his blood infection.

Component Results

Component	Value	Flag	Reference Range	Status
WBC	76.2	HH	3.5-11 K/uL	Corrected
Comment: ADJUSTED FOR NUCLEATED RBC'S CORRECTED ON 09/13 AT 1553: PREVIOUSLY REPORTED AS 79.2				
RBC	3.74	L	4.47-5.91 M/uL	Final
Comment: Critical Value reported to and read back by HOWES BY RD AT 1608 FOR WBC ON 091309				
Hemoglobin	12.6	L	13.5-17.5 g/dL	Final
Hematocrit	35.2	L	41-53 %	Final
MCV	94.0		81-99 fL	Final
MCH	33.6	H	26-33 pg	Final
MCHC	35.8	H	32-35 g/dL	Final
RBC Dist Width	15.5	H	<14.3 %	Final
Platelet Count	180		150-450 K/uL	Final
Mean Platelet Volume	8.1		6.7-11.4 FL	Final

Figure 11F: Blood analysis of pre-mortem blood in reference to the normal range

Consistent with the reduction of the red blood cell count, there is a significant increase in iron level coupled with the hemolytic degradation of the heme (deep red iron-containing) components (Figure 11G). These data suggested that there was hemolytic activity associated with the destruction of the red blood cells in Malcolm's pre-mortem blood. *Streptococcus pyogenes* could have been the organism introduced to Malcolm's blood to provide the hemolytic destruction of his red blood cells.

Special Studies, Reports, and Consultations
Saved serum from pre mortem studies was used for additional post mortem clinical
chemistry testing. These studies showed:

Serum ferritin 392.530 ng/ml (normal 20-300)
Total iron binding capacity 648mcg/dl (normal 230-430)
Iron 541 mcg/dl (normal 40-160)

Figure 11G: Iron Content in Malcolm's pre-mortem blood

According to Malcolm's medical history, total iron in his pre-mortem blood had increased significantly to 543 mcg/dl (micrograms per deciliter) in contrast to 40–160 for the normal range.

It was unclear, though, how much of the total iron came from the ruptured red blood cells and how much came from Malcolm's genetic disposition of hemochromatosis (a condition that causes the body to retain iron in human tissue; see Part IX below).

VII. If Streptococcus Had Not Killed Him, Plague Would Have Killed Him

The abundance of iron released from oxygen-laden hemoglobin carrier molecules complements the growth of *Yersinia pestis* and other organisms, which means they flourish in the presence of high iron content in the blood. So it was obvious from Malcolm's autopsy and medical history that *Streptococcus* played a helper role to facilitate the growth of KIM D27-Cm^{R} in his blood and cause his plague death.

Schneewind had told us on September 18, 2009, that two other bacterial species were present in Malcolm's contaminated blood.

Malcolm had not one but two deadly bacterial species. So if *Streptococcus* had not killed him, plague would have killed him.

VIII. How Was the Streptococcus Bacteria Introduced to His Blood?

My second question was how *Streptococcus pyogenes* had gotten into Malcolm's blood—just in time for their hemolytic function to complement the plague growth. This could not be a coincidence. Unlike *Y. pestis*, *Streptococcus* belongs to a family of bacteria that do not thrive in blood and is not a blood-borne pathogen by nature. There is only one explanation that could allow this to happen. *Streptococcus* was planted in the same infection mix brought to his blood by intragastric delivery, the same mechanism that delivered plague bacteria to his blood.

Since most Streptococcus families of bacteria do not thrive in blood and not a blood borne pathogen by nature, few molecules that multiplied in the blood formed the Nutritional Variant Streptococcus (NVS), a spin-off satellite colony, detectable in Malcolm's pre-mortem blood.

specimen. The Gram negative rods were later identified as an attenuated Yersinia pestis strain (KIM D27), and the Gram positive cocci were later identified as nutritionally variant streptococci. There was

Figure 11H: Nutritional Variant Streptococci was the predominant form of Streptococcus found in Malcolm's blood

In reviewing all of the evidence thus far, I believe Malcolm had been given the *E. coli* lysate mix on August 31, 2009, and the infection mix the next day on September 1, 2009, in accordance with the bioterrorism protocol Part I (Chapter 10, Table 1).

Per Schneewind and Chervonsky's bioterrorism protocol, part I of the infection mix consists of *E. coli* DH5-alpha that has been boiled in water and mixed with streptomycin and orally delivered to the victim on day 1. The purpose is to allow *E. coli* to trap villous M cells embedded in the intestinal epithelium to create lesions that will allow other organisms to enter the bloodstream (see Chapter 10, Figure 10C). Part 2 of the infection mix consists of organisms that can enter through lesions created by trapped villous M cell and transverse across basement membrane to the bloodstream and replicate in blood and cause a fatal infection.

The *Streptococcus* bacteria that brought death to Malcolm, I believe, were delivered mechanistically via the two-day infection protocol. The hemolytic activity associated with *Streptococcus* bacteria was meant to destroy the oxygen machinery in the blood, causing increased difficulty in breathing and eventually suffocation and death as in Malcolm's case.

IX. Generalized Delivery Of Organisms to Blood

Per Schneewind and Chervonsky's protocol, the infection mix is not limited to the delivery of plague bacteria, any bacteria that can replicate in blood or those with limited replicating potential can be target delivered to blood and still cause fatal death.

The *Streptococcus* bacteria, in this instance, were not a natural blood borne pathogen, had not caused a blood infection in the wild. It is not known if Streptococcus replicated productively in Malcolm's blood, if so, to what extent. But the hemolytic activity associated with the rupture of the RBC was real and caused significant debilitation of the breathing machinery and death. This may explain the limited presence of the helper bacteria but predominantly the Nutritional Variant Streptococcus variant that was in Malcolm's blood as noted in the Autopsy Report.

It was nonetheless a generalized delivery protocol devised by Schneewind and Chervonsky to have brought in the Streptococcus bacteria and caused disruption of the breathing machinery that led to Malcolm's death. It was indeed an award winning protocol that perfected the Bioterrorism. Malcolm had indeed become a tester for this generalized protocol delivery. The University of Chicago had become the ARRA award recipient three days after the mystery death that took Malcolm's life in a storm.

I felt sick at heart. I asked, *just what sins had Malcolm committed that he deserved such horrible treatment.* And Malcolm alone was the chosen martyr for bioterrorism. He alone paid its price with his life.

X. Hemochromatosis—Iron Overload

The autopsy report further revealed a genetic disposition that Malcolm unknowingly carried. A hereditary hemochromatosis, a genetic disease characterized by an overload of iron in the blood and organs of the affected individuals. DNA sequencing of his HFE genetic locus identified a C282Y mutation of both of his chromosome 6 at location 6p21.3, proving that Malcolm was indeed a homozygote in HFE mutation (that is, he had the mutation on both genes and therefore expressed the mutation, rather than just being a carrier of it). C282Y designated a shift of amino acid change from cysteine to a tyrosine at the 282nd amino acid in the HFE locus, and the change affected the ability of the HFE protein to transport iron. Because of the mutation, iron in the body of a hemochromatosis individual gets accumulated in tissue and blood, which results in many phenotypic symptoms associated with the disease. (A vast literature on the genetic and metabolic disease of hemochromatosis is available online.)

XI. Hemochromatosis Is Not the Cause of Death

Because of the defect in pgm^- strain, KIM D27 normally would require iron transport for its multiplication and metabolism in peripheral tissue. That is, extra iron would be required to allow KIM D27 replication. But in the blood of a normal healthy individual, there is significant amount of iron present, 40–160 mcg/dl (Figure 11G), to support KIM D27 multiplication. Therefore KIM D27, albeit iron requirement, can be virulent—cause death—in the blood of normal healthy individuals as well as in that of hemochromatosis individuals.

Schneewind presented at the NIH-RAC meeting on June 16,2010, (http://videocast.nih.gov/Summary.asp?File=16009) in which he noted that there were at least twenty-two metabolic pathways where iron could be scavenged from the blood cell environment to enable KIM D27 multiplication. KIM D27 infection in the human blood is like a lethal injection as Schneewind repeatedly told NIH at the NIH-RAC meeting. The excess iron from a hemochromatosis individual may aid KIM D27 multiplication in peripheral tissues, but in iron-rich blood environment, KIM D27 can effectively multiply as its wild type counterpart. This was in agreement with what Schneewind noted in his NIH Agent Profile form, "untreated pneumonic and septicemic are fatal." (Figure 9F). Therefore, hemochromatosis or iron overload did not contribute to the cause of Malcolm's death.

The actual cause of death was the portal of entry that allowed the deadly bacteria to gain access to blood. And in Malcolm's case, "the presence of focally ulcerated mucosal prolapse in the colon, the portal of entry through the gastrointestinal tract" was the cause of death.

Nevertheless, iron had become the subject of a university alibi that did not make sense to anyone except the infectious disease experts and federal investigators who preached the doctrine for one reason—a cover-up.

XII. Hereditary Hemochromatosis Restores Virulence in blood?

Schneewind published another report in 2012 in the *Journal of Infectious Disease* to demonstrate that the hereditary hemochromatosis restores the virulence of plague vaccine strain in knockout mice (http://www.ncbi.nlm.nih.gov/pubmed/22896664).

In this experiment, a line of transgenic knockout mice had been generated employing the genetic mutation that changed at the amino acid of C282Y in the HFE locus. This line of mouse strain was created to mimic Malcolm's genetic disposition to see if his clinical isolate of KIM D27-Cm^R (UC91309) could restore virulence and cause the knockout mice to die as Malcolm did.

This report is problematic and it is based on false assumptions. First, the experiment was done based on subcutaneous inoculation (injection beneath the skin). Malcolm did not, as the autopsy clearly stated: there was no sign of subcutaneous inoculation. Malcolm received the agents that killed him through an oral portal of entry and intragastric route of delivery. That makes his infection consistent with the oral ingestion portal of entry and gastrointestinal route of infection as outlined in the bioterrorism protocol discussed in Chapter 10.

Second, Malcolm's pre-mortem isolate, KIM D27-Cm^R, later called UC91309, was HIPPA protected and could not be used without HIPPA release.

Malcolm had been murdered. Schneewind and his cohorts had used his blood isolate to conceal the true cause of death. Therefore this article by Quenee and others (2012) had no scientific value in so far as it has no bearing on how Malcolm was killed and the true cause of death.

The Haunting Truth

I felt devastated by the truth revealed in the autopsy report, the medical record, and the FOIA documents. The truth of Malcolm's death—that it had to have been intentionally planned and executed—and the excruciating way in which he died, all the pain he endured in his last hours, haunted me. For days, weeks, and months afterward, I often woke from nightmares, my body drenched in sweats and tears running down my face.

M. H. Pappworth, a nineteenth-century British physician who pulled back the dark curtain on human experimentation, once wrote,

> "No doctor has the right to choose martyrs for science or for the general good" (*Human Guinea Pigs*).

Malcolm certainly did not choose his fate.

12

CONFESSION, OBSTRUCTION AND SILENCE

It was almost Christmas 2009. Schneewind and his Microbiology Department were in spirits and ready to celebrate the fantastic year they had just had. Four days after Malcolm's death, President Obama had announced a stimulus award of $42 million from the American Recovery and Reinvestment Act (ARRA) to the University of Chicago and $1 million to Schneewind's lab (Figure 10R).

In the glow of candlelight, Christmas trees, alcohol, and caviar, Schneewind talked about the death of a scientist in his lab, a sacrifice in the cause, advancing the science of biological warfare, proving a protocol. Swayed by the highball in his hand, Schneewind delivered his ghastly tale to his department in a chillingly upbeat voice. Shock immobilized his listeners. It was as if a mission had just been completed, and he reveled in the handsome payoff of the ARRA award.

He went further then, comparing Malcolm's death to that of the late Howard Taylor Ricketts, after whom the National Containment Laboratory was named. Was this a furtherance of the fantastical tale that Malcolm was somehow complicit in his own death? Or a twisted idea that it was all right to sacrifice someone else's life in the pursuit of science?

Howard Taylor Ricketts

In the early 1900s, Ricketts had been a physician, a faculty member in the University of Chicago's Department of Pathology and Bacteriology, and a dedicated researcher. According to a fellow physician and close friend of his, Ricketts, while researching one disease, needed a human subject and so injected himself with cultures (T. W. Goodspeed, "Howard Taylor Ricketts," *The University Record*, April 1922: 105–106). Although he grew ill, Ricketts recovered from that experience, and then went on to make important discoveries in the origins of Rocky Mountain spotted fever and later typhus fever. Sadly, he died of typhus in 1910 while in Mexico investigating the origins.

And now, almost 100 years after his death, another scientist at the university had died of a deadly disease. The culprit was KIM D27-Cm^R, a derivative that specified chloramphenicol resistance to KIM D27 parent. The acquisition of chloramphenicol resistance to KIM D27 was engineered by Malcolm; then, designated by Schneewind to be used in the bioterrorism human trial on October 7, 2008. The isolate recovered from Malcolm's pre-mortem blood revealed the source and nature of his infection: Malcolm was besieged by the Bioterrorism weaponry. (See Chapter 10, Bioterrorism Warfare Development: Part III: Human Trials using KIM D27-Cm^R)

Somehow, it seemed to listeners, Schneewind found a certain satisfaction in this, as if it were an appropriate way to commemorate Ricketts's death.

People who had not been at the Christmas party heard whispered retellings of it; the story eventually spread to other academic institutions, including Harvard and MIT, Malcolm's *alma mater*. I heard a report of that macabre evening much later, sitting in a hotel room with someone who had been there. My informer spoke quietly, and I heard the revulsion in his voice as well as the incredulity, still, that it had happened.

Some who heard the story suggested a lawsuit against the perpetrators. But that was much easier said than done, as the family would discover. There were more people, at least more powerful people, invested in covering up the truth than there were those who wanted it exposed.

Many people simply could not believe that in twenty-first-century America, there were those in a respected university who would carry out experiments on human subjects without conscience or the subjects' consent.

Experimentation Without Consent or Conscience

Nazi doctors experimented without conscience or remorse on those helpless to resist them. In Nuremburg, in 1947, a group of them faced trial for their actions. When the military tribunal passed judgment on the doctors, it also laid out what is now known as the Nuremburg Code. The code captures many of what are now taken to be the basic principles governing the ethical conduct of research involving human subjects. The first principle:

> The voluntary consent of the human subject *is absolutely essential*.
>
> This means that the person involved should have legal capacity to give consent; should be so situated as to be able to exercise free power of choice, without the intervention of any element of force, fraud, deceit, duress, over-reaching, or other ulterior form of constraint or coercion...[Emphasis added]

Born in Germany, with an M.D. and a Ph.D. from the University of Cologne, Schneewind grew up and was educated under the shadow of Nazi's atrocities and in the bright light of the Nuremburg Code. He had been honored in the United States by being named the Louis Block professor and chair of the University of Chicago's Microbiology Department and the principle investigator of the esteemed GLRCE. Yet despite the highest honors and the horrifying lessons of the past, this man could evidently reduce his colleagues, researchers and students alike, to mere laboratory animals. He could claim to be protecting them with a vaccine while in reality destroying them—at least one of them—with an agent of biological warfare. Did students and researchers in Chicago's bioterrorism laboratory know what was pumped into their blood had the potential to transmit devastating diseases and even death?

Investigation Hijacked

On February 12, 2010, Schneewind, along with Krista Currell, an advocate for risk and patient safety for the University of Chicago, talked to the family over the telephone. To our surprise, Kathy Ritger was not included in the conversation, though she was the chief investigator. Despite his Christmas confession, Schneewind stuck to his original story. Currell was there to "put the nails into Malcolm's coffin." She parroted the story, as told to federal agencies, that Malcolm seldom wore gloves and lacked the necessary training in biosafety standards because he either rarely attended the biosafety training courses or was always late. In other words, she and Schneewind asserted that Malcolm's death was a result of his own negligence and noncompliance.

No one from the family replied to this—and no one believed Currell or Schneewind. As our lawyer had advised, we merely listened, not giving voice to our suspicions and anger. Hearing the unscrupulous and defamatory remarks against Malcolm, the family lost complete confidence in the University of Chicago.

What about the confession that came in the candlelight in the midst of caviar and highballs? Was everyone at Chicago a fool? And what about the threat of vengeful retaliation against my children?

When I spoke to Kathy Ritger the following week, she told me in confidence that she did not want to be coached by Schneewind on what she would be allowed to say and not to say. So she had excused herself from the conference call, saying that she was moving at that time and not available.

I believe that Kathy Ritger soon found out about the Christmas confession. She must have felt betrayed in the investigation in which she promised to be fair and neutral. In our conversations in the following weeks, she no longer dwelt on any subjects related to her investigation, much less showed an interest in investigating the diabetic needles that she talked energetically about soon after Malcolm's death. Her attitude changed from being critical to a sense of frustration, even helplessness, of no longer being in charge of where the investigation was going. It seemed like she was being shutdown in her efforts to uncover the truth. I could tell that Kathy had completely lost her respect for Schneewind, his Microbiology Department, and the institutions that she was investigating.

Death Threat

Someone did not like the questions I was asking. Some time after the Christmas confession at the Microbiology Department, all the falsehoods about Malcolm, and the calls from Schneewind and Currell and Kathy Ritger, I received a death threat.

Some friends and I had a trip to the National Parks in the next state planned. Before we left, one of my friends received a call from an informant warning that we might be killed on the road in an "accident." The informant advised my friends not to drive their own car.

My friends did not tell me about the threat immediately. I only knew that the decision had been made to rent a car for the trip. The second day on our trip, my friends were reluctant to travel farther and decided we should return home immediately. After we reached home safely, they told me about the call and their fear that we would all be killed.

Fear, horror and disbelief—all enveloped me. Were we talking about a university and respected professors or a Chicago gang and its members, who struck out to intimidate, punish, and silence?

Later that month, I communicated this event to the NIAID and the person who had helped me with the FOIA requests. She sounded as shocked and disbelieving as I had been and wanted me to report the incident to authorities. But the Chicago police had been satisfied to go along with the federal and state investigators' determination that nothing criminal had taken place. In Incident Reporting to NIH-OBA on October 16, 2009, and later on November 23, 2010, Kanabrocki relayed the following to NIH-OBA (Figure 12B).

> The Chicago Police Department reported that there was no evidence of criminal activity. The CDPH, IDPH and CDC concluded that this incident appeared to be an isolated incident bearing no threat to the public and that no systemic issues at the UC were identified. OSHA and NIOSH concluded their investigation issuing no citations or fines to the UC.

To whom should I turn to report my death threat?

The Chicago Police Department reported that there was no evidence of criminal activity. The CDPH, IDPH and CDC concluded that this incident appeared to be an isolated incident bearing no threat to the public and that no systemic issues at the UC were identified. OSHA and NIOSH concluded their investigation issuing no citations or fines to the UC.

Figure 12B: Police report of the Death incidence

Obstructers, Accessories, and Silence

There was a pervasive sense that people were being silenced as the investigation continued, affecting not only Kathy Ritger but also others at the university and the various agencies involved. Thinking of all of this, I decided one night to look at the federal laws governing conduct in an investigation.

- *Obstruction of Proceedings Before Departments, Agencies, and Committees (18 U.S. Code § 1505)*

  Whoever corruptly . . . influences, obstructs, or impedes or endeavors to influence, obstruct, or impede the due and proper administration of the law under which any pending proceeding is being had before any department or agency of the United States . . . [s]hall be fined under this title, imprisoned not more than 5 years or, if the offense involves international or domestic terrorism . . . , imprisoned not more than 8 years, or both.

- *Obstruction of Criminal Investigation (18 U.S. Code § 1510)*

  Whoever willfully endeavors by means of bribery to obstruct, delay, or prevent the communication of information relating to a violation of any criminal statute of the United States by any person to a criminal investigator shall be fined not more than $5,000, or imprisoned not more than five years, or both.

- *Accessory After the Fact (18 U.S. Code § 3)*

  Whoever, knowing that an offense against the United States had been committed, receives, relieves, comforts or assists the offender in order to hinder or

> prevent his apprehension, trial or punishment, is an accessory after the fact.

One University of Chicago faculty member expressed regret over the treatment of Malcolm—and me—in a letter to the family.

> [I]t is likely that many of the faculty at the University of Chicago feel an uneasy sense of guilt at the treatment both your father but especially your mother received here. I've never really discussed it with any of my colleagues but I get the sense that we think that although everything may have been legally correct it was somehow wrong. [Letter to family, in possession of author]

Silence over what happened in my case did not concern me now. But silence over what was done to Malcolm did—that was much, much worse.

*****

Albert Einstein, who fled Nazi Germany and came to the United States for a safer and more just life, once said, "If I were to remain silent, I'd be guilty of complicity."

*****

The Reverend Martin Luther King Jr., who gave his life in the struggles for civil rights, preached, "He who passively accepts evil is as much involved in it as he who helps to perpetrate it."

*****

American president John F. Kennedy, who also sought to change the world for the better, also warned us, "The only thing necessary for the triumph of evil is that good men do nothing."

*****

13

THE GRAND COVER-UP

Schneewind's confession from the Christmas party certainly shocked a lot of people. Among them were members of the faculties and staff and scientists who were concerned about the criminal activity going on at the university. Some felt injustice, some felt regret. Some felt exultation for the mission completed. It was the wildest game they played. First came murder, then came confessions, and then the elaborate cover-ups. As Ling had said, "Dead man cannot talk. It is time to move on to help the ones still living."

Isn't it justice in America that murderers must pay for their crimes? Only if there is the will to investigate the crime and make the perpetrators pay. That will seemed lacking in Malcolm's death. Officials from the university and various federal agencies gathered to repair the reputation of the University of Chicago and cover up the crime by positing a "perfect storm."

OBA, RAC, and Business as Usual

One of the primary responsibilities of the Office of Biotechnology Activities is overseeing *dual use research of concern* (DURC), which is defined as:

> life sciences research that, based on current understanding, can be reasonably anticipated to provide knowledge, information, products, or technologies that could be directly misapplied to pose a significant threat with broad potential consequences to public health and safety, agricultural crops and other plants, animals, the environment, materiel, or national security. The United States Government's oversight of DURC is aimed at preserving the benefits of life sciences research while minimizing the risk of misuse of the knowledge, information, products, or technologies provided by such research. [http://osp.od.nih.gov/office-biotechnology-activities/biosecurity/dual-use-research-concern]

Dual use research means research that can be used to develop biological warfare agents as well as vaccines—research that can be used to kill as well as protect. NIH-OBA had the responsibility to make sure DURC was not being misused, yet those who worked for OBA seemed more concerned with ensuring that it was business as usual at the University of Chicago than with investigating the dark truth behind Malcolm's death.

In May 2010, there was a flurry of email exchanges between Joseph Kanabrocki, UC's assistant dean for biosafety, and Jacqueline Corrigan-Curay, the acting director of OBA, about an upcoming Recombinant DNA Advisory Committee (RAC) meeting (Figure 13A-G). The central issue concerned getting the funding for the bioterrorism grant to the University of Chicago, which had been put on hold since Malcolm's death. At the same time, it was necessary to determine if there should be policy changes in the protocol for handling attenuated strains of *Y. pestis* in light of Malcolm's death. The relevant policy issues were to be discussed at the June 2010 RAC meeting.

On May 13, 2010, NIH-OBA's Acting Director Corrigan-Curay sent UC's Kanabrocki the proposed notice about the issue to be discussed at the RAC meeting and commented on the reference in it to Malcolm's death:

> "Joe
>
> Will be trying to publish this FR [Federal Register] notice shortly. Please see the reference to the death at U of Chic. I have made it deliberately vague regarding the strain but just wanted to give your legal folks a heads up." (Emphasis added, Figure 13A)

Quickly, Kanabrocki replied that he had forwarded the draft announcement to the university general counsel [GC]:

> "Thanks Jacqueline,
>
> I have forwarded the draft announcement to our GC for their review. For what it is worth, I agree with you that the reference to UC contains only information that is already in the public realm. I will get back to you as soon as I hear back from the Office of GC.
>
> I appreciate the heads-up." (Emphasis added, Figure 13B)

Corrigan-Curay urged action soon. "Please see if they can live with that one reference". Then she commented:

> "I think the call went well. One difficulty is that there seems to be agreement that there is a threshold that an organism have the potential to impact public health but there is also some uneasiness about not having a say regarding how many antibiotics you put into it" (Emphasis added, Figure 13C).

It would appear that NIH-RAC meeting participants had a general concern about:

- Threshold of virulence with the pgm^- [pigmentation-deficient] strains (not able to say KIM D27).
- How many antibiotic resistances it specified by the pgm^- strains.

The concerns raised by the RAC meeting participants were those discussed in the biological warfare agent protocol that Schneewind had submitted to NIH-OBA (Chapters 9 and 10). I am sure that Corrigan-Curay had seen the pathogenic profile of the KIM D27 and its fatality rate of 100 percent for septicemic infection (Figure 9F). The acquisition of chloramphenicol resistance by KIM D27-Cm^R strain was used in the human trial (Chapter 10, Part III).

On Sunday, May 16, 2010, Kanabrocki replied to Corrigan-Curay:

> "After several rounds of email the past three days, our GC [General Counsel] is finally OK with *the reference given that it is coming from NIH and not UC.* Thanks very much for the opportunity to get their read [Emphasis added].
>
> I agree that the discussion went well. I look forward to additional discussions to arrive at a paradigm that will hopefully be consistent and generally applicable." (Emphasis added, Figure 13D)

In reply, Corrigan-Curay on May 18, 2010 agreed to cover up the true identity of the strain that infected Malcolm,

> "OBA can state at the mtg that it is not the pgm^- strain if you are not able to say that as we have that information and it is relevant to the work of the RAC." (Emphasis added, Figure 13E)

Mission accomplished, on May 19, 2010, Kanabrocki reaffirmed the university's request to have this information come from NIH and he himself would confirm on the sideline should it be necessary.

> "Thank you Jacqueline.
>
> Since this detail was included in the report sent by UC to OBA, this information is already in the public realm and is not, in any case, HIPPA protected, at least from my perspective. That said, I believe it would be best to have this info come from you, if relevant to the discussion. I will, of course, confirm this detail in the meeting should it be necessary.
>
> I appreciate your sensitivity and consideration of the UC dilemma as it relates to this discussion. I continue to work diligently on my end to facilitate to full disclosure of details related to this LAI case. In my opinion, these details have direct relevance to this discussion and will provide an important perspective when considering the disease potential of attenuated *Y.p.* strains." (Emphasis added, Figure 13F)

The university and the NIH were clearly working together to return the bioterrorism grant award to the university after it had been withheld due to Malcolm's death. They were prepared to tell the RAC meeting participants, the scientists at large, and the general public whatever it took and to hide the critical health issues raised by Malcolm's death.

The HIPPA nondisclosure was used as an excuse by the university not to answer questions in public disclosure raised by the RAC meeting participants. It also prevented the family from knowing the details surrounding Malcolm's death and headed off a potential lawsuit against the university and certain of its scientists for Malcolm's wrongful death.

****

From: Corrigan-Curay, Jacqueline (NIH/OD) [E] [mailto:corrigaja@od.nih.gov]
Sent: Thursday, May 13, 2010 10:47 AM
To: Kanabrocki, Joseph [BSD] - MIC
Subject: ProposedAction_Transfer_Chloramphenicol_resistance_LLNL_draft_JCC_err_051110.doc

Joe
Will be trying to publish this FR notice shortly. Please see the reference to the death at U of Chic. I have made it deliberately vague regarding the strain but just wanted to give your legal folks a heads up.

Thanks
Jacqueline

Figure 13A: Collaborated cover-up of NIH-OBA with Kanabrocki, university biosafety Dean

****

From: Kanabrocki, Joseph [BSD] - MIC [mailto:jkanabro@bsd.uchicago.edu]
Sent: Thursday, May 13, 2010 12:02 PM
To: Corrigan-Curay, Jacqueline (NIH/OD) [E]
Subject: RE: ProposedAction_Transfer_Chloramphenicol_resistance_LLNL_draft_JCC_err_051110.doc

Thanks Jacqueline,

I have forwarded the draft announcement to our GC for their review. For what it is worth, I agree with you that the reference to UC contains only information that is already in the public realm. I will get back to you as soon as I hear back from the Office of GC.

I appreciate the heads-up.

-joek

Joseph Kanabrocki, Ph.D., C.B.S.P.
Assistant Dean for Biosafety
Associate Professor of Microbiology

Figure 13B: collaborated cover-ups between Kanabrocki and Corrigan-Curay

****

On May 16, 2010, at 6:50 PM, "Corrigan-Curay, Jacqueline (NIH/OD) [E]" <corrigaja@od.nih.gov> wrote:

Joe
I need to get the FR notice out on Monday. Please see if they can live with that one reference

I think the call went well. One difficulty is that there seems to be agreement that there is a threshold that an organism have the potential to impact public health but there is also some uneasiness about not having a say regarding how many antibiotics you put into it
Thanks
Jacqueline

Jacqueline Corrigan-Curay, J.D., M.D.
Acting Director
Office of Biotechnology Activities
National Institutes of Health
(301) 496-9838
(301) 496-9839 fax
corrigaja@od.nih.gov

Figure 13C: collaborated cover-ups between university Kanabrocki and NIH-OBA, Corrigan-Curay

****

From: Kanabrocki, Joseph [BSD] - MIC [mailto:jkanabro@bsd.uchicago.edu]
Sent: Sunday, May 16, 2010 8:35 PM
To: Corrigan-Curay, Jacqueline (NIH/OD) [E]
Subject: Re: ProposedAction_Transfer_Chloramphenicol_resistance_LLNL_draft_JCC_err_051110.doc

Dear Jackie,

After several rounds of email the past three days, our GC is finally OK with the reference given that it is coming from NIH and not UC. Thanks very much for the opportunity to get their read.

I agree that the discussion went well. I look forward to additional discussions to arrive at a paradigm that will hopefully be consistent and generally applicable.

Best wishes
-joek

Sent from my iPhone

Figure 13D: University Counsel agreed with the reference used and specified its direction from NIH

****

From: Corrigan-Curay, Jacqueline (NIH/OD) [E] [mailto:corrigaja@od.nih.gov]
Sent: Tuesday, May 18, 2010 12:57 PM
To: Kanabrocki, Joseph [BSD] - MIC
Subject: RE: ProposedAction_Transfer_Chloramphenicol_resistance_LLNL_draft_JCC_err_051110.doc

Joe
OBA can state at the mtg that it is not the pgm- strain if you are not able to say that as we have that information and it is relevant to the work of the RAC
Thanks
jcc

Jacqueline Corrigan-Curay, J.D., M.D.
Acting Director
Office of Biotechnology Activities
National Institutes of Health
(301) 496-9838
(301) 496-9839 fax
corrigaja@od.nih.gov

Figure 13E: Corrigan-Curay on the cover-up with Kanabrocki over the strain issue.

****

From: Kanabrocki, Joseph [BSD] - MIC [jkanabro@bsd.uchicago.edu]
Sent: Wednesday, May 19, 2010 11:34 AM
To: Corrigan-Curay, Jacqueline (NIH/OD) [E]
Subject: RE: ProposedAction_Transfer_Chloramphenicol_resistance_LLNL_draft_JCC_err_051110.doc

Thank you Jacqueline.

Since this detail was included in the report sent by UC to OBA, this information is already in the public realm and is not, in any case, HIPAA protected, at least from my perspective. That said, I believe it would be best to have this info come from you, if relevant to the discussion. I will, of course, confirm this detail in the meeting should it be necessary.

I appreciate your sensitivity and consideration of the UC dilemma as it relates to this discussion. I continue to work diligently on my end to facilitate to full disclosure of details related to this LAI case. In my opinion, these details have direct relevance to this discussion and will provide an important perspective when considering the disease potential of attenuated *Y.p.* strains.

Best wishes,
-joek

Figure 13F: Cover-ups and fabrication by Kanabrocki and NIH regarding Malcolm's death

NIH-RAC finally met on June 16, 2010, and reviewed "evidences regarding the ability of the strain to cause disease" based on derailed and fabricated data sheets. The information hidden from the reviewers were:

- KIM D27's pathogenic profile (Figure 9F)
- Chloramphenicol Antibiotic resistance added to the KIM D27 strain (Figure 10J, 10K, 10L).
- KIM D27-Cm^R recovered from the deceased (Figure 10M, 10N)
- The portal of entry and route of infection that caused Malcolm's septicemic plague and death. (Figure 11B, Chapter 10—Bioterrorism Protocol, Part I, II, III, IV and V)

Memorandum to NIH RAC Committee – 6/15/2010

A memorandum dated June 15, 2010 was sent to the RAC participants to convince them that researchers should be able to continue studying and manipulating attenuated strains of *Yersinia pestis* without the oversight and biosafety measures required for "deadly strains." (Figure 13G). The memorandum contained the following erroneous assertions, assertions that left members with the impression that these strains were not deadly.

I. **Statement in 3(a) from Memorandum**

   "Pgm^- mutants possessing all the remaining virulence determinants of *Y. pestis* were isolated and shown to be *avirulent*." (Emphasis added)

   Response to Statement

   Not true as discussed in the following sources

   - $LD_{50}=10^7$

- Titball and Williamson 2004. Review of *Y. pestis* plague vaccines, including the statement: "there are significant concerns over the safety of this vaccine [using EV76]. The vaccination of mice or vervets with this strain can result in fatalities" (see Chapter 9—"Attenuated Strains: EV76 and Human Trials" and Figure 9F)
- Quenee et al. 2008, including this statement: "mice immunized with 1 × 10^7 CFU of *Y. pestis* KIM D27 presented clinical symptoms (ruffled fur and lethargy) and 30% mortality" (see Chapter 7, "Plague—Live Attenuated Vaccine Strain, KIM D27" and Figures 7A and 7B).

II. Statement in 3(a) from Memorandum

"Subsequent studies demonstrated that *pgm–* mutants are completely *avirulent* in mice infected from peripheral routes (*subcutaneous* or peritoneal injection)." (Emphasis added)

Response to Statement

Not true as discussed in the following sources

- 50% fatality (see Chapter 9, Figure 9F)
- Quenee et al. 2008 (see Chapter 7, "Plague—Live Attenuated Vaccine Strain, KIM D27" and Figure 7B)
- Titball and Williamson 2004 (see Chapter 9, "Attenuated Strains: EV76 and Human Trials" and Figure 9F)

III. Statement in 3(a) from Memorandum

"The fact that virulence is conditional has provided a biological containment that *protects laboratory workers from contracting plague while allowing studies of pathogenesis in models of systemic (septicemic) plague*." (Emphasis added.)

Response to Statement

- Animal studies that mimic septicemia death at 100% fatality (see Chapter 7, Figure 7C)
- University of Chicago, Select Agent Institutional Biosafety Committee, Agent Profile Form for *Yersinia pestis* KIM D27, "untreated pneumonic and septicemic [plague] are fatal" (see Chapter 9, Figure 9G)
- It is a bioterrorism warfare in development—septicemic infection and death (Chapter 10, Parts I, II, III and IV)
- On August 27, 2011, a second researcher working in Schneewind's laboratory at the University of Chicago was infected by an organism under study. Although she lived, thanks to prompt treatment with surgery and antibiotics, assumptions about safety in the laboratory, particularly Schneewind's laboratory, were clearly wrong. (See below for further discussion of this incident.) (see http://news.sciencemag.org/2011/09/updated-university-chicago-microbiologist-infected-possible-lab-accident?ref=hp)

IV. Statement in 3(c) from Memorandum

"Y. pestis Strain, EV 76, is pgm⁻ and was used for large-scale *subcutaneous* vaccination in humans in Madagascar without the '*slightest mishap.*'" (Emphasis added, Figure 13G)

Response to Statement

Not true as discussed in the following sources

- Titball and Williamson 2004 (see Chapter 9, "Attenuated Strains: EV76 and Human Trials" and Figure 9F)
- Debora MacKenzie 2009 (See Figure 9E)

V. Statement in 3(f) from Memorandum

Regarding Malcolm's death, "[w]e don't know what manipulations were performed with the organism or what biosafety practices were used."

Response to Statement

- Oral ingestion/intragastric delivery was hidden and replaced with "subcutaneous." Bioterrorism protocol (Chapter 10, Part I-V, and Chapters 11, 12).
- NIH-OBA had the details of Malcolm's death on file but chose to conceal them from the committee members who drafted the memorandum. In the absence of any hard facts, the cause of death by the committee members had been derailed to that of "breach of biosafety" and "lack of training" for Malcolm's death.
- Malcolm died from the biological warfare agent given to him orally that caused his septicemic death (Chapters 7–11). His death was marked as poison on a note deposited in the FDA CBER Office by members of the federal investigation team.

As the cover-up got under way, Kanabrocki was satisfied that his tale went smoothly with his friends from NIH, CDC, FDA, and OBA.

> "By the way, on Friday I indeed did learn that an MMWR report if forthcoming, I received a draft of this report [the MMWR draft]. The senior author is Dr. Kathy Ritger from the Chicago Department of Public Health. Marty Schriefer is also an author." (Figure 13H).

June 15, 2010

MEMORANDUM

TO: The National Institutes of Health
Recombinant DNA Advisory Committee

FROM: Members of the Community of
Plague Researchers including:

a. Pgm⁻ mutants possessing all the remaining virulence determinants of *Y. pestis* were isolated and shown to be avirulent (6, 11). Subsequent studies demonstrated that *pgm* deletion (*Δpgm*) mutants are completely avirulent in mice infected from peripheral routes (subcutaneous or peritoneal injection); however, such mutants are virulent when introduced intravenously in mice (2, 17). The fact that virulence is conditional has provided a biological containment that protects laboratory workers from contracting plague while allowing studies of pathogenesis in models of systemic (septicemic) plague.

c. *Y. pestis* Strain E.V. (or various substrains such as EV 76) is Pgm^- and was used for large-scale subcutaneous vaccination of humans in Madagascar without "the slightest mishap" (13, 14, 19).

We are aware that plague researcher Dr. Malcolm Casadaban recently died from laboratory-acquired systemic disease due to a *Δpgm* Lcr^+ *Y. pestis* strain. Dr. Casadaban had hereditary hemochromatosis, which would be equivalent to iron overload, which is known to permit virulence from a peripheral route of infection in experimental animals. We don't know what manipulations were performed with the organism or what biosafety practices were used. The incident illustrates that if a researcher breaches the host barriers that protect against disease from *Δpgm Y. pestis*, he/she can get serious disease. However, the incident does not alter the fact that plague researchers with normal defenses have worked with *Δpgm* Lcr^+ *Y. pestis* for decades without adverse consequence.

Figure 13G: Excerpts from Memorandum to NIH-RAC committee members to NIH RAC

Kanabrocki became the assistant dean for biosafety in 2007 when he came to the University of Chicago. It was around the same time Malcolm was demoted from a tenured professor to a research associate and later a "laboratory worker" in Schneewind's laboratory.

These actions were taken and communications made behind closed doors in NIH and were kept from the family of the deceased.

From: Kanabrocki, Joseph [BSD] - MIC [mailto:jkanabro@bsd.uchicago.edu]
Sent: Monday, June 21, 2010 6:59 PM
To: Corrigan-Curay, Jacqueline (NIH/OD) [E]
Subject: RAC Y.p. discussion

Dear Jacqueline,

I have two questions related to the RAC *Y.p.* discussion last week.

First, is the *Major Actions* exemption for chloramphenicol use limited to the *pgm*⁻ strains or does it also extend to the fully virulent strains. I suspect it is limited to the attenuated strains but my mind is fuzzy about that part of the motion since we were focused on the *pgm*⁻ strains.

Second, the motion specifically called out chloramplenicol use as being exempt from RAC review as a *Major Action* for the *pgm*⁻ strains; but what about other markers commonly used even in fully virulent strains, such as ampicillin and kanamycin? I expect that the use of these markers in *pgm*⁻ strains would NOT be considered a *Major Action*, but again, it was not clear to me from my recollection of the motion wording. I believe this was the point that Dr. Schneewind was trying to make.

By the way, on Friday I indeed did learn that an MMWR report is forthcoming; I received a draft of this report. The senior author is Dr. Kathy Ritger from the Chicago Department of Public Health. Marty Schreifer is also an author. Of course, I will forward it to you as soon as I learn of its publication. It will be informative.

Thanks again for your help with all of this. I thought that we had a very good discussion.

Best wishes,
-joek

Joseph Kanabrocki, Ph.D., C.B.S.P.
Assistant Dean for Biosafety
Associate Professor of Microbiology
Biological Sciences Division
University of Chicago
920 E. 58th Street, CLSC 705A
Chicago, IL 60637
(773) 834-7496 (Hyde Park Campus)
(630) 252-2390 (Ricketts Lab at Argonne)
(773) 612-6804 (mobile)

Figure 13H: Kanabrocki received a draft of the MMWR on Malcolm's death

The scale of the cover-ups went far beyond what anyone could have imagined. NIH and its National Labs fostered in the university setting intended to build the "biological weaponry" in the Chicago laboratory and to conceal the real facts of Malcolm's death from the RAC attendants. It was "the development of warfare agents," rather than protection of the health of the general public, that those in the federal government and the university that were involved cared about.

On July 6, 2010, Louis Kirchhoff from the NIH-RAC meeting had concerns about the pgm^{-} strain that was delivered to Malcolm. Corrigan-Curay replied and stated about "subQ or intramuscular" delivery for his cause of death (Figure 13J). This, of course, was not how Malcolm had been infected.

> "Thanks Louis. I wonder if it has to do with the route of exposure. The vaccine being subQ or intramuscular and the potential that the researcher had an pulmonary exposure, although we do not know. Certainly, in the animal studies route of exposure was critical." (Emphasis added, Figure 13J)

From: Corrigan-Curay, Jacqueline (NIH/OD) [E] [mailto:corrigaja@od.nih.gov]
Sent: Tuesday, July 06, 2010 2:40 PM
To: 'Kirchhoff, Louis'
Cc: Kanabrocki, Joseph [BSD] - MIC
Subject: RE: Yersinia pestis pgm-

Thanks Louis. I wonder if it has to do with the route of exposure. The vaccine being subQ or intramuscular and the potential that the researcher had an pulmonary exposure, although we do not know. Certainly, in the animal studies route of exposure was critical.
Jacqueline

Figure 13J: In reply to Kerchhoff's questions how the vaccine was delivered to Malcolm, Corrigan-Curay replied as "subQ" and "intramuscular".

NIH-RAC passed the consensus vote to reinstate the research on the biological warfare agent in GLRCE in the aftermath of Malcolm's death.

Undoubtedly, this provision made the university and NIH very happy. Research went forward as before at GLRCE, protocols unchanged. Malcolm alone was the sacrifice.

It was appalling to see federal agencies engaged in the cover-up of foul play, especially in light of the US statutes governing federal investigators in Title 18 U.S.C. § 3; § 1505 and § 1510 that address the accessory after the fact, corruption, and obstruction of criminal investigation (see Chapter 12).

On June 22, 2010, Corrigan-Curay forwarded the draft of the MMWR to Roizman, Kanabrocki, Kirchhoff, and others by email (Figure 13K). Curiously at the bottom of her email, she noted that she was including "an article from Dr. Collins and Dr. Hamburg that may be of interest."

Could this be a hint that the draft MMWR actually came from Dr. Collins, NIH director, and Dr. Hamburg, FDA commissioner? I knew Francis Collins had been kept aware of all activities involved with Malcolm's death in early emails (Figure 8B, 8C) and Margaret Hamburg, a bioterrorism specialist appointed by President Obama in 2009. Could this be the NIH that used billions of dollars of US taxpayer money to fund the bioterrorism project (Figure 10R) that ended with a murder and then a grand cover-up?

NIH Draft of the MMWR

The draft of the *Morbidity and Mortality Weekly Report* (MMWR) about Malcolm's death could not have been any further from the truth. It brought out a new issue that I had not contemplated from before. "Interviews with coworkers identified inconsistencies in the deceased's biosafety practices." Francis Collins and Margaret Hamburg had used these allegations to assign responsibility of the Chicago crime to the deceased himself.

The NIH draft MMWR reported,

> "This is the first reported laboratory acquired infection and fatality caused by attenuated *Y. pestis.* Histopathology and immunohistochemistry indicate septicemic, not pneumonic plague indicating Percutaneous or mucocutaneous exposure" (Emphasis added, Figure 13K)

Even more appalling were the mocking words,

> "Although mouse inoculation studies confirms that the infecting strain was attenuated, hemochromatosis-induced iron overload might have contributed to host susceptibility by creating an environment conductive to pathogenesis" (Emphasis added, Figure 13K).

I finally understood what Kanabrocki was talking about, "our GC [General Counsel] is finally OK *with the reference given that it is coming from NIH and not UC*" (Figure 13D). And "That said, I believe, *it would be best to have this info come from you*, if relevant to discussion" (Figure 13F, emphasis added).

From: Corrigan-Curay, Jacqueline (NIH/OD) [E] [mailto:corrigaja@od.nih.gov]
Sent: 06/22/2010 11:11 a.m.
To: (Bernard.Roizman@bsd.uchicago.edu); BartletJ (Jeffrey.Bartlett@NationwideChildrens.org); Borror, Kristina C (HHS/OPHS); David Williams; Dr. Buchmeier; Dr. Fong; Dr. Mastroianna; Dr. Ross; Dr. Wei; Ertl (ertl@wistar.org); Fan, Hung ; 'Hawkins1@georgetown.edu'; Howard_Federoff (hjf8@georgetown.edu); james_yankaskas; 'Jeffkahn@umn.edu'; John Zaia; Joseph Kanabrocki; 'kodishe@ccf.org'; Kirchhoff, Louis; mmallino@aol.com; 'sjflint@princeton.edu'; 'sstrome@smail.umaryland.edu'; Takefman, Daniel M (FDA)
Cc: Groesch, Mary (NIH/OD) [E]; Shipp, Allan (NIH/OD) [E]; Gargiulo, Linda (NIH/OD) [C]; Jambou, Robert (NIH/OD) [E]; Montgomery, Maureen (NIH/OD) [E]; O'Reilly, Marina (NIH/OD) [E]; Rosenthal, Eugene (NIH/OD) [E]; Shih, Tom (NIH/OD) [E]; Siddiqui, Mona (NIH/OD) [C]
Subject: Yersinia and Personalized Medicine - but not together

Dear RAC members:

Joe Kanabrocki was sent the following link after our meeting so apparently there is confirmation that the research who died had hemachromatosis. I am not sure that would change the conclusion regarding having antibiotic markers placed into Pgm- strains analyzed under III-A-1. Obviously, more research needs to be done.

I have also included an article from Dr. Collins and Dr. Hamburg that may be of interest.

Thank you for all of the hard work you put in at the meeting last week.

Sincerely,

Jacqueline

Fatal Case of Laboratory-Acquired Infection with an Attenuated *Yersinia pestis* Strain of Plague — Illinois, 2009

AUTHORS: ***Andrew Medina-Marino, M. Schriefer, S. Black, P. Mead, K. Weaver, K. Metzger, B. King, S. Gerber, W-J. Shieh, S. Zaki, S. Cali, C. Conover, K. Soyemi, K. Ritger***

BACKGROUND: In September 2009, a researcher working with KIM-D27, *a Yersinia pestis* strain attenuated by deletion of iron-acquisition genes, died of acute septicemia. *Y. pestis*, the cause of plague, was isolated from blood cultures. We investigated the source of infection, strain virulence, and contributing host factors.

METHODS: We conducted an environmental assessment, interviewed laboratory personnel, and reviewed autopsy and medical records. Laboratory investigation included histopathologic and immunohistochemical analysis of autopsy samples, genetic testing, plasmid DNA profiling, and polymerase chain reaction (PCR)-based characterization of the strain isolated from the deceased; virulence studies were conducted in mice.

RESULTS: No deficiencies were identified in required laboratory maintenance. Interviews with coworkers identified inconsistencies in the deceased's biosafety practices. Immunohistochemistry revealed *Y. pestis* within blood vessels of all organs but not alveolar airspaces. Histopathology identified abnormal liver iron deposits. Markedly elevated pre-mortem serum iron levels were identified postmortem. Genetic testing confirmed hereditary hemochromatosis, an iron-overload disease. Plasmid DNA and PCR analysis identified the infecting strain as KIM-D27. Of mice inoculated with KIM-D27 stock strain, 2/64 (3%) died versus 0/64 inoculated with the deceased's strain (P=.15); 19/24 (79%) mice inoculated with unattenuated *Y. pestis* died.

CONCLUSIONS: This is the first reported laboratory-acquired infection and fatality caused by attenuated *Y. pestis*. Histopathology and immunohistochemistry indicate septicemic, not pneumonic plague, indicating percutaneous or mucocutaneous exposure. Although mouse inoculation studies confirm that the infecting strain was attenuated, hemochromatosis-induced iron overload might have contributed to host susceptibility by creating an environment conducive to pathogenesis. Studies to assess the pathologic contribution of iron overload are being conducted in hemochromatosis-mutant mice. Hemochromatosis might represent a new risk factor for infections with bacteria attenuated by iron-acquisition defects.

KEYWORDS: plague, laboratory-acquired infections, hemochromatosis, risk factor

Figure 13K: NIH draft of the MMWR

M.D., Ph.D., and J.D.

The article shared deception and fabrication in the criminal activities of a murder conjured in grand cover-up, sealed by Francis Collins and Margaret Hamburg. I couldn't help noticing the level of stellar scholarship associated with the perpetrators and investigators: Schneewind M.D., Ph.D.; Corrigan-Curay M.D., J.D.; Francis Collins, M.D., Ph.D., Margaret Hamburg, M.D., Karen Frank, M.D., Ph.D., Wun-Ju Shieh, M.D., Ph.D., and many more. I was intimated by such an array of doctors, scientists, lawyers and federal investigators. But I was not afraid because the truth had spoken. Indeed, the Truth had spoken.

The major players at the university were happy. Ken Alexander, Chief of Pediatric Infectious Diseases, could promote the whole commotion as a "perfect storm," with nothing changed, when he spoke to the *Chicago Maroon* newspaper on April 8, 2011. Alexander's statements not only diminished and disrespected Malcolm but also mocked him. "I'm sure that if Dr. Casadaban had had one comment for us as we sat around that table it was: 'Listen guys, I'm trying to teach you something, and you better damn well learn it.' " Alexander said, "and I think we did."

After the NIH-RAC meeting and forwarding of the MMWR draft, the mission was accomplished as planned. The University of Chicago and NIH, through a joint effort, satisfactorily appeased everyone in the larger scientific community. It was time to let the world know how Malcolm had died, as least according to the story jointly generated by NIH and the University of Chicago. It was time to tell the fabricated story to the world. Several major publications started to appear and to note the cause of death in Malcolm after that. Of course, these contained the misinformation that had been promoted to cover up Malcolm's true cause of death. These reports are summarized in the following section.

I. MMWR by CDC

"Fatal Laboratory-Acquired Infection with an Attenuated *Yersinia pestis* Strain, Chicago, Illinois, 2009" MMWR by CDC

(http://www.cdc.gov/mmwr/preview/mmwrhtml/mm6007a1.htm?s_cid=mm6007a1_w).

Reported by: K Ritger, M.D., et al. on February 25, 2011

II. New England Journal of Medicine

"Investigation of a Researcher's Death Due to Septicemic Plague"

(http://www.nejm.org/doi/full/10.1056/NEJMc1010939?viewType=Print&)

Reported by: Frank, Schneewind and Shieh on June 30, 2011

III. Chicago Maroon

"Plague death uncovered" Chicago Maroon

(http://chicagomaroon.com/2011/04/08/plague-death-uncovered/)
Reported by Jingwen Hu on April 8, 2011

- "This report summarizes the results of that investigation, which determined that the cause of death likely was an unrecognized occupational exposure (route unknown) to *Y. pestis,* leading to septic shock." ----See Chapter 10, part V.
- "Hemochromatosis-induced iron overload might have provided the infection by KIM D27 strain, which is attenuated as a result of defects in its ability to acquire iron" ----See Chapter 11, part XI
- "Researchers should adhere to recommended biosafety practices when handling any live bacterial cultures, even attenuated strains." ---See Chapter 13.
- "On September 10, he notified his 'supervisor' about his illness to explain his absence from work. Whether the patient himself suspected his symptoms were consistent with plague is not known" ---Malcolm reported "flu-like symptoms" as in *Biosafety*

Manual for Working with Yersinia pestis strain KIM D27. Malcolm was an associate professor using facility in the laboratory of Schneewind to conduct his NIH awarded experiments. In this report, the writer had downgraded Malcolm's academic status and Schneewind became the supervisor.

- "A review of attendance records for university biosafety training identified deficiencies in staff attendance (including the patient) at a number of required biosafety courses." ---See IBC minutes in Figure 8M.
- "CDC also identified a chloramphenicol resistance gene (a common, laboratory-based resistance marker) that was not in the original laboratory stock strain, suggesting that the infecting strain had been modified as part of routine laboratory research." ---See bioterrorism protocol, part III).
- "Male Swiss Webster mice were inoculated *subcutaneously* with varying doses of bacteria ranging from 10^3 to 10^8 colony-forming units (CFUs)." ---Incorrect portal of entry and route of infection contradictory to the autopsy report.
- "Analysis of the UC91309 genome sequence revealed the insertion of an antibiotic-resistance cassette that had been engineered by the patient as a research activity, indicating that the patient isolate was the laboratory-manipulated strain and not a naturally occurring U.S. strain." ---See bioterrorism protocol, Part III of Chapter 10.
- "The route of entry of the organism was not determined despite a thorough autopsy and an investigation." ---See autopsy report.
- "It is possible that inadvertent exposure to a subcutaneous or mucous membrane had occurred." ---A deadly pathogen deliberately target delivered to Malcolm's blood via oral, intragastric delivery.
- "Still, how the bacterium got into his blood remains a mystery." ---False, read this book.
- "When you deal with an attenuated organism . . . they are attenuated for most people but not for everybody," said Dr. Ken

Alexander, Infectious Disease at the U of C pediatrics department. ---False. Untreated septicemic plagues are fatal, universally fatal.

- "Still, how the bacterium got into his blood remains a mystery. According to the report, Casadaban did not attend all of the required classes and did not consistently wear gloves while handing Y. pestis." ---False, the mystery was in Chapter 10.
- According to Alexander, "this constituted a breach of safety protocol, but it's unclear whether it contributed to his death."---False, Kanabrocki conjured this in his dream.
- The University has made no changes to laboratory procedures in response to Casadaban's death. The circumstances surrounding Casadaban's death comprised a "perfect storm," and policies are not changed based on an aberration, Alexander said. ---see this book.

Second Infection in Schneewind's Laboratory

Two years after Malcolm died, another scientist conducting research in Schneewind's laboratory got infected. This time it involved a microbiologist who contracted *Bacillus cereus*, supposedly through skin contact, possibly after encountering inoculant spilled by another researcher. Yet could this story be taken at face value given what had been done to Malcolm and the truth of his death? This time, the victim was rescued and sent to the hospital for treatment on August 27, 2011. She, of course, was held to blame for her own infection.
(*http://www.councilforresponsiblegenetics.org/blog/post/University-of-Chicago-Microbiologist-Infected-From-Possible-Lab-Accident.aspx.)*

At this point, I had no respect for the federal agencies or the university involved. Great institutions such as the University of Chicago, NIH, and DOD seemed not to care about what really happened to the scientists in the house of bioterrorism. How convenient it is to blame the sick and the deceased for the bioterrorism project that delivered poison and death now that all voices for truth had been silenced.

Kanabrocki's Accusations Against Malcolm

On September 24, 2012, Kanabrocki did a presentation "Apparent Violations Warranting Citations Were Not Found During the Inspection Period." In the following slides, Kanabrocki boldly proclaimed that Malcolm's death was self-inflicted because "training records revealed deficiencies in attendance of required training by some lab staff (included deceased)" and "interviews with co-workers revealed inconsistent compliance by deceased with lab PPE policies." In another, Kanabrocki said, "Inconsistent use of gloves could have resulted in an inadvertent transdermal exposures."

In all, Kanabrocki, the dean of biosafety, painted a grotesquely distorted picture of Malcolm as an entry-level "laboratory worker." Kanabrocki clearly colluded with Schneewind to reduce researchers, lab staff, and even faculties in his department to guinea pigs, and Malcolm to a careless lab worker who had caused his own death. Together with Schneewind, he committed murder.

Code of Conduct

Kanabrocki seized every opportunity to characterize Malcolm's death as self-inflicted, due to noncompliance and inconsistency in wearing gloves in a RG2 laboratory. It was easy to attack Malcolm now, when he could no longer defend himself.

As his "lesson learned" in his slides, Kanabrocki established a Code of Conduct to be practiced by all investigators in the Microbiology Department to report on peers when sloppy laboratory techniques were observed. Further, he raised the issue of moral and ethical conduct and scientific integrity. Had he forgotten the bioterrorism protocol and infamous human trials with deadly pathogens and the oral delivery that induced septicemic death in a murder trail that could shake the earth and make every decent human cry? The accusations against Malcolm, whom I knew to be a meticulous and careful researcher when working with pathogens, send a deep, burning anger through my veins.

In his biosafety lecture in 2013, Kanabrocki asked for written pledge of Code of Conduct from all members of the Microbiology Department. This turned a murder case into a self-inflicted death that Kanabrocki had orchestrated at the university (Figure 13L). http://mrce.wustl.edu/mrce/ckfinder/userfiles/files/MRCE%20Biosafety_LAI_Sept_2013.pdf "Biosafety_LAI_Sept_2013 copy.pdf"

A written *Code of Conduct* has been established and each investigator in the Dept. of Microbiology has signed this agreement.

- Includes provisions for peer reporting, self reporting.
- Includes commitment to adherence to established lab safety standards.
- Includes provisions for ethical conduct and scientific integrity.

Figure 13L: A Code of Conduct to mock the Deceased by Kanabrocki in lecture.

Alas! This is the group of scientists who populated the Department of Microbiology at the University of Chicago, who committed murder, confessed to the murder, and fabricated a cover-up with various federal officials so they could continue their research into the area of biological warfare without hindrance or real oversight.

Months later, I came across the Schneewind's "Code of Conduct" (http://oba.od.nih.gov/biosecurity/meetings/Jan2011/Schneewind_Olaf.pdf). There appeared a man who was a Dr. Jekyll and Mr. Hyde for the twenty-first century, professing to care about the safety of those working in his laboratories and developing way to combat bioterrorism, on the one hand, while refining the weapons of biological warfare and colluding in the death of a colleague on the other. What kind of man, a Louis Block professor, could do this? What kind of university could see it done? Was Malcolm's death really a "perfect storm"? Or a play orchestrated by the elements of the government and the university in the name of bioterrorism?

14

HUMAN SACRIFICE UNDER U.S. BIOTERRORISM AND U.S. LAW

What kind of nation is the United States? How important are the rights of the individual? Are all people truly seen as being equal? Is life sacrosanct here? As Americans, we like to view ourselves as the guys in the white hats, the white knights, and defenders of those who cannot defend themselves.

But our past, even our most immediate past, is checkered. We have not always done the right thing, and in many cases we have let the wrong thing be done. The hats that some Americans have worn have been very black indeed.

U.S. Laws Sanctioning Human Experimentation

Two laws, which were only repealed in the late 1990s after they came to public awareness, exemplify just how wrong U.S. governmental policy and public officials have been in the past:

- Public Law 95-79, Title VIII, Sec. 808, July 30, 1977, 91 Stat. 334.(In U.S. Statutes-at-Large, Vol. 91, page 334, you will find Public Law 95-79.)

> "The use of human subjects will be allowed for the testing of chemical and biological agents by the U.S. DOD, accounting to Congressional committees with respect to the experiments and studies."

- Public Law 97-375, Title II, Sec. 203(a)(1), Dec. 21, 1982, 96 Stat. 1882. (In U.S. Statutes-at-Large, Vol. 96, page 1882, you will find Public Law 97-375.)

> "The Secretary of Defense [may] conduct tests and experiments involving the use of chemical and biological [warfare] agents on civilian populations [within the United States]."

Although these laws were repealed or modified in late 1990s, the fact that they were ever the law of the land is frightening. They specifically allowed the U.S. government the right to experiment on its citizens without their knowledge or consent. And the people doing those experiments were doctors, the ones who swore *primum non-nocere*—"first, do no harm."

Unintentional harm—through a wrong diagnosis or failure to discover an underlying health problem immediately, for instance—can be bad enough but doctors are only human and make mistakes. Intentional harm, including that done under the guise of scientific research for the supposed "betterment" of humankind, is an unforgiveable crime. Shouldn't we have learned that truth at Nuremberg, after World War II, when all the horrors of Nazi medical experimentation were revealed in the Doctors Trial?

One result of the trial, in addition to the conviction of sixteen physicians, was the Nuremberg Code. This code, developed by the Nuremberg Military Tribunal, offered standards by which human experimentation could be conducted. The code captures many of what are now taken to be the basic principles governing the ethical conduct of research involving human subjects. In a review of the Nuremberg Code fifty years after it was articulated, scientist Evelyne Shuster wrote:

> "The judges at Nuremberg, although they realized the importance of Hippocratic ethics and the maxim *primum non-nocere,* recognized that more was necessary to protect human research subjects. Accordingly, the judges articulated a sophisticated set of 10 research principles centered not on the physician but on the research subject. These principles, which we know as the Nuremberg Code, included a new, comprehensive, and absolute requirement of informed consent (principle 1), and a new right of the subject to withdraw from participation in an experiment (principle 9)."

Principles 1 and 9 have too often been ignored, despite the horrors brought forth during the Doctors Trial. In the infamous Tuskegee Syphilis Study, in the United States, hundreds of black men with syphilis were promised free medical care if they would agree to go to doctors at the Public Health Service. In a CDC 2013 article, a writer described the study:

> The study was conducted without the benefit of patients' informed consent. Researchers told the men they were being treated for "bad blood," a local term used to describe several ailments, including syphilis, anemia, and fatigue. In truth, they did not receive the proper treatment needed to cure their illness. In exchange for taking part in the study, the men received free medical exams, free meals, and burial insurance. [CDC, 2013, "U.S. Public Health Service Syphilis Study at Tuskegee," Timeline, http://www.cdc.gov/tuskegee/timeline.htm]

Despite the lessons of Nazi Germany and the Nuremberg Code, the Tuskegee medical researchers continued their study for *forty years*. Not until the Associated Press broke the story in 1972 did the government end the study. What kind of doctors could callously watch men under their care suffer and die from a disease that could have been treated with penicillin? How could those doctors live with themselves, when they had taken oaths to heal?

They were just steps away from being the same kind of men as those who infected Malcolm and left him to die a horrible death, slowly suffocating for hours until his body gave out. Those doctors knew the Nuremberg Code yet it meant no more to them than Malcolm's life.

Nuremberg Code

Here are the ten principles of the Nuremberg Code.

1. The voluntary consent of the human subject is absolutely essential.

   This means that the person involved should have legal capacity to give consent; should be so situated as to be able to exercise free power of choice, without the intervention of any element of force, fraud, deceit, duress, over-reaching, or other ulterior form of constraint or coercion; and should have sufficient knowledge and comprehension of the elements of the subject matter involved as to enable him to make an understanding and enlightened decision. This latter element requires that before the acceptance of an affirmative decision by the experimental subject there should be made known to him the nature, duration, and purpose of the experiment; the method and means by which it is to be conducted; all inconveniences and hazards reasonably to be expected; and the effects upon his health or person which may possibly come from his participation in the experiment.

   The duty and responsibility for ascertaining the quality of the consent rests upon each individual who initiates, directs or engages in the experiment. It is a personal duty and responsibility, which may not be delegated to another with impunity.

2. The experiment should be such as to yield fruitful results for the good of society, unprocurable by other methods or means of study, and not random and unnecessary in nature.

3. The experiment should be so designed and based on the results of animal experimentation and knowledge of the natural history of the disease or other problem under study that the anticipated results will justify the performance of the experiment.

4. The experiment should be so conducted as to avoid all unnecessary physical and mental suffering and injury.

5. No experiment should be conducted, where there is an *a priori* reason to believe that death or disabling injury will occur; except, perhaps, in those experiments where the experimental physicians also serve as subjects.

6. The degree of risk to be taken should never exceed that determined by the humanitarian importance of the problem to be solved by the experiment.

7. Proper preparations should be made and adequate facilities provided to protect the experimental subject against even remote possibilities of injury, disability, or death.

8. The experiment should be conducted only by scientifically qualified persons. The highest degree of skill and care should be required through all stages of the experiment of those who conduct or engage in the experiment.

9. During the course of the experiment, the human subject should be at liberty to bring the experiment to an end, if he has reached the physical or mental state where continuation of the experiment seems to him to be impossible.

10. During the course of the experiment the scientist in charge must be prepared to terminate the experiment at any stage, if he

> has probably cause to believe, in the exercise of the good faith, superior skill and careful judgment required of him that a continuation of the experiment is likely to result in injury, disability, or death to the experimental subject.

Human Experimentation in History

The following chronology on human experiments comes from several sources, including the Alliance for Human Research Protection (AHRP), an organization whose "mission is to stand up—and speak out—for the human rights of research subjects—especially those who are vulnerable and/or susceptible to coercion, manipulation and exploitation." ("About the Alliance for Human Research Protection," http://www.ahrp.org/cms/content/view/18/87/, accessed April 17, 2014.) Some of the experiments were in other countries; some right here in the United States. That such evil could go on in the names of science and medicine sickens the soul and makes one despair for the future of humankind.

Timeline of Torture

The Alliance for Human Research Protection and other human rights watchdogs and advocates have complied the following list of atrocities:

> **1936–1945:** Japan establishes a compound at Pingfan, an occupied area of China, and puts Dr. Shiro Ishii in charge of what would be known as Unit 731. Ishii begins tests of germ warfare and vivisection experiments on thousands of Chinese soldiers and civilians. Some Russian prisoners may also have been victims of Ishii's atrocities. Americans numbered among his victims as well, a fact that would not come to light until decades later, along with the horrific details of Ishii's crimes against humanity.

Prisoners were usually dissected while still alive. Neither sex nor age protected victims from the scalpels of Ishii and his doctors, who experimented on babies and pregnant women (often impregnated by the doctors themselves) as well as on other women and men. One Japanese medical researcher later recalled, "I cut [the prisoner] open from the chest to the stomach and he screamed terribly and his face was all twisted in agony. He made this unimaginable sound, he was screaming so horribly. But then finally he stopped. This was all in a day's work for the surgeons, but it really left an impression on me because it was my first time."

1942: U.S. Army and Navy doctors infect nearly 400 prisoners in Chicago with malaria in order to study the effects of new experimental drugs to combat the disease. No one informed the prisoners about the nature of the experiments. Tellingly, Nazi doctors later on trial at Nuremberg cite this American study to defend their own actions during the Holocaust.

1944–1946: University of Chicago Medical School professor Dr. Alf Alving conducts malaria experiments on more than hundreds of Illinois prisoners.

1945: Manhattan Project doctors inject plutonium into three patients at Billings Hospital at the University of Chicago to study the effects.

1945: Project Paperclip is initiated. The U.S. State Department, U.S. Army intelligence, and the CIA recruit Nazi scientists and offer them immunity and secret identities in exchange for work on top-secret government projects in the United States.

1946: The U.S. government offers Ishii and other Unit 731 leaders a secret deal. The United States will cover up their atrocities and give them immunity from war crimes prosecution in exchange for the biowarfare agent data based on human experimentation. According to a top-secret U.S. Army Far East Command report: "The value to the U.S. of Japanese biological weapons data is of such importance to national security as to far outweigh the value accruing from war-crimes prosecution." This is verified by 1956 FBI memorandum. The U.S. government placed a higher value on knowledge gained than on the lives of human beings and on justice for those tortured by doctors.

1946–1974: The Atomic Energy Commission authorized a series of experiments in which radioactive materials are given to individuals, many of whom were not informed that they were the subjects of an experiment. Subjects were selected from vulnerable populations such as the poor, elderly, and mentally retarded children (who were fed radioactive oatmeal without the consent of their parents), and also from students at University of California–San Francisco. In 1993, the experiments were uncovered and made public. In 1996, the United States settled with the survivors for $4.9 million.

1986: A report to Congress reveals that the U.S. government's current generation of biological agents includes modified viruses, naturally occurring toxins, and agents that are altered through genetic engineering to change immunological character and prevent treatment by all existing vaccines.

1987: DOD admits that, despite a treaty banning research and development of biological agents, it continues to operate research facilities at 127 facilities and universities around the nation.

1990: The FDA grants DOD waiver of Nuremberg Code for use of approved drugs and vaccines in Desert Shield.

1994: Senator John D. Rockefeller issues a report revealing that for at least fifty years the DOD has used hundreds of thousands of military personnel in human experiments and for intentional exposure to dangerous substances. Materials included mustard and nerve gas, ionizing radiation, psychochemical, hallucinogens, and drugs used during the Gulf War.

1996: DOD admits that Desert Storm soldiers were exposed to chemical agents.

1997: Eighty-eight members of Congress sign a letter demanding an investigation into bio-weapons use and Gulf War Syndrome.

Timeline of Proposed Protection Against Human Experimentation

There have been attempts to protect people from secret and nonconsensual experimentation. (This shorter listed has been gathered by the same human rights advocates and watchdogs.)

1947: The Nuremberg Code is laid out by the military tribunal when it passes sentence on the Nazi doctors in the infamous Doctors Trial.

1964: World Medical Association adopts Helsinki Declaration, asserting, "The interests of science and society should never take precedence over the well being of the subject."

1966: U.S. National Institutes of Health's Office for Protection of Research Subjects ("OPRR") created and issues Policies for the Protection of Human Subjects calling for establishment of independent review bodies later known as Institutional Review Boards (IRB).

1975: HHS promulgates Title 45 of Federal Regulations titled "Protection of Human Subjects," requiring appointment and utilization of IRBs.

1990s: U.S. laws allowing secret and nonconsensual human experimentation are repealed or modified.

Victims from the View of Medical Experimenters

How can anyone, but especially doctors, experiment on people without conscience or remorse of any kind? The answer is that the experimenters do not view their victims as fellow human beings but as inferior creatures—like the rats and other animals they inflict with diseases.

Guatemala Syphilis

Syphilis was a major health threat in 1940, causing blindness, insanity and even death. Many of the same researchers had carried out studies in Guatemala and on prisoners in Terre Haute, Indiana, but unlike the Guatemalan patients, the Americans gave consent. For years, the experiments were secret, until a medical historian at Wellesley College, Susan Reverby, in Massachusetts found the records among the papers of Dr. John Cutler, who led the experiments.

A federal commission of bioethics was set up to learn more at the disclosure of the Guatemala syphilis case. According to Markel, ethical considerations in science began to emerge after World War II, and further enlightenment followed after the American civil rights movement. "This was far too common a phenomenon until our recent history—in the prison population and homes for the mentally retarded," he said. "Part of the reason we did this research is we didn't think of them as humans."

Nazi Doctors

The discovery of the Holocaust and the murder of 6 million people—Jews, the disabled, homosexuals, and gypsies, as well as bad experimentation—by Nazi doctors, opened the world's eyes. The founding of the United Nations and the World Health Organization also brought attention to human rights. "Each new discovery and advance in social rights, we had to learn the lesson over and over again," said an observer. "For a long time, blacks were second or third class citizens."

Quotes from Advocates to End Human Experimentation

Many people have spoken out against human experimentation. Among them:

Mike Stobbe, Associated Press: "AP IMPACT: Past Medical Testing on Humans Revealed," *Washington Post*, Sunday, February 27, 2011 (http://www.washingtonpost.com/wp-dyn/content/article/2011/02/27/AR2011022700988.html)

Shocking as it may seem, U.S. government doctors once thought it was fine to experiment on disabled people and prison inmates. Such experiments included giving hepatitis to mental patients in Connecticut, squirting a pandemic flu virus up the noses of prisoners in Maryland, and injecting cancer cells into chronically ill people at a New York hospital.

U.S. officials also acknowledged there had been dozens of similar experiments in the United States—studies that often involved making healthy people sick.

Inevitably, they will be compared to the well-known Tuskegee syphilis study. In that episode, U.S. health officials tracked 600 black men in Alabama who already had syphilis but didn't give them adequate treatment even after penicillin became available.

> "These studies were worse in at least one respect—they violated the concept of "first do no harm," a fundamental medical principle that stretches back centuries. "

****

Authur Caplan, director of the University of Pennsylvania's Center for Bioethics:

> "When you give somebody a disease—even by the standards of their time—you really cross the key ethical norm of the profession."

****

M. H. Pappworth, nineteen-century British physician and author of *Human Guinea Pigs*:

> "No doctor has the right to choose martyrs for science or for the general good. "

****

World Medical Association, *Declaration of Helsinki: Ethical Principles For Medical Research Involving Human Subjects*:

> "In medical research on human subjects, considerations related to the well being of the human subject should take precedence over the interests of science and society."

****

Barack Obama, president of the United States, at Prayer Breakfast on February 6, 2014:

> "[K]illing the innocent [is the] ultimate betrayal of God's will".

****

Justice Edward J. Greenfield, in T.D. v New York State Office of Mental Health 626 N.Y.S.2d 1015 (1995):

> "The mere mention of experimental medical research on incapacitated human beings—the mentally ill, the profoundly retarded, and minor children summons up visceral reactions with recollections of the brutal Nazi experimentation with helpless subjects in concentration camps, and elicits shudders of revulsion

> when parallels are suggested. Even without the planned brutality, we have had deplorable instances of overreaching medical research in this country. "

****

Michael Ellner, president of Health Educations AIDS Liaison, an advocacy group for HIV parents, referring to experimental treatment at Manhattan's Incarnation Children's Center [http://www.survivreausida.net/article5922.html]:

> "They are torturing these kids, and it is nothing short of murder. "

****

Francis Boyle, professor and author of his book *Biowarfare and Terrorism*:

> "The University . . . and its handpicked "Ethics Committees" had a vested economic interest in approving unethical if not illegal scientific "research" contracts to be conducted on campus that involve scientific experimentation on animals and human beings. I suspect the same is true at most other University campuses around the country today. The same principle holds true for the federally mandated IBCs at American Universities, private biotech companies, and U.S. Government labs that are supposed to supervise the safety of recombinant-DNA research projects."

****

Authors of the *Review of the Tuskegee Syphilis Study*:

> "Society can no longer afford to leave the balancing of individual rights against scientific progress to the scientific community. "

Anyone Can Become a Victim

Malcolm was not a member of an underprivileged class of citizens; he was a faculty member at the University of Chicago, an acknowledged genius by many of his peers. He advocated against the human experimentation program at the university, where dangerous biological weapons, under the guise of vaccines, were tested on laboratory researchers. He fought the systems and the processes and discussed his concerns openly. As Ken Alexander told the *New York Times* on September 21, 2009, Malcolm "hoped to create a better vaccine for plague in part because of concerns about its possible use in bioterrorism."

The circumstances of Malcolm's death and all that I found out because of it, including the sordid history of medical research involving involuntary human subjects, persuaded me that major reforms need to be implemented to restrain U.S. bioterrorism research and policy. My recommendations would include:

1. U.S. laws allowing any type of human experimentation should be scrutinized to make sure that all people are informed and protected. Laws giving anyone the right to experiment on other human beings should be challenged and reviewed by the U.S. Supreme Court.
2. Have a whistle blower blog set up for victims and families to come forward.
3. Identify all human subjects who fell victims to the current illegal vaccine projects.
4. Open a forum to allow victims and their families to present their grievances to the U.S. Administration.
5. Educate the judicial system to allow victims and their families to seek legal remedies and justification.
6. Allow compensation system under current law to help victims and dependents.

15

THE TRUTH SHALL SET YOU FREE

Malcolm Casadaban did not die as the result of an accidental infection. He was deliberately infected with the plague using a distinctive two-day infection protocol created in a national laboratory tasked, by the federal government, with developing biowarfare protocols and products. His death required a highly orchestrated effort to target deliver the deadly pathogen to his blood and ensure his death.

Malcolm's death marked a continuation of government-sanctioned experimentation on unsuspecting and non-consenting citizens, a practice seen time and again in U.S. history, in programs such as the Tuskegee experiment and the testing of biowarfare products on U.S. armed forces. Federal and state agencies and officials, including individuals at the University of Chicago, conspired to cover up the true circumstances of Malcolm's death. Any investigator who raised the question of foul play was silenced. This allowed conspirators to fabricate the story of a "careless researcher" who helped destroy himself in "the perfect storm," which placated the public and allowed the scientists to continue business as usual. The truth remained hidden; the murderers remained unpunished. There was no justice for Malcolm, who suffered not only assassination of his person but also assassination of his character.

To find justice for Malcolm, I set out on my own journey of investigation. It was a long and arduous road. There were officials who refused to talk to me and agencies that claimed they had no information for me (when I knew that they did). I was threatened with death. What I did find out gave me nightmares, as I understood better how horrific Malcolm's last days and hours had been, how much he had suffered as he fought for each labored breath. Yet no matter how much evidence I found, no one wanted to listen, to pursue the truth, to sue or prosecute the criminals. Still I have persevered for the sake of justice, for personal integrity, and for love, love for the gentle and generous soul I had been married to for twenty-eight years, the devoted father of my children, my best and truest friend.

By telling Malcolm's story, I hope to shed light not only on the crime against him, carried out by "respected" scholars and scientists, but on similar crimes perpetrated in this country and elsewhere in the name of scientific, medical, or political advancement. They are, in reality, acts of bioterrorism.

> Biological weapons may be employed in various ways to gain a strategic or tactical advantage over an adversary, either by threats or by actual deployment . . .. These agents may be lethal or non-lethal, and may be targeted against a single individual, a group of people, or even an entire population. They may be developed, acquired, stockpiled or deployed by nation states or by non-national groups. In the latter case, or if a nation-states uses it clandestinely, it may also be considered bioterrorism. [Microbiology, 2013, p. 1306]

All acts of bioterrorism, whether they target a nation or a single person, are acts against humanity, are acts against everything we believe to be just and right and decent. When we as individuals and as a country allow the innocent like Malcolm to be tortured and sacrificed without protest, we open the door to evil and to the ultimate destruction of our society and our very souls.

When Malcolm died, the world lost a renowned and innovative scientist. It also lost a kind, caring, and giving human being. Other people have spoken of his scientific achievements. I want to finish this book by speaking of his achievements as a person and, by doing so, showing just how much we lost when he died.

Brooke Remembers Her Father

On September 16, 2009, Brooke delivered her eulogy for her father.

> Most of you refer to him as "Malcolm" or even "Mal" or "Professor." But to Leigh and me, he was always "Dad."
>
> He came from a big southern family, born and raised in New Orleans. Although Leigh and I grew up here in Chicago, every summer and Christmas was spent in New Orleans with the rest of the large family. Family was very important to my father.
>
> He had such a big heart. He enjoyed the company of others, he was passionate about his work, and most of all he loved his family. He cared for Leigh and me dearly.
>
> My childhood memories of Dad include our weekend camping trips, the season pass to Six Flags, and bike rides along the coast. While words cannot exactly describe what my childhood was like being raised by a genius, I can honestly say that my dad provided us with a wonderful life. I love you, Dad.

Leigh Remembers Her Father

On September 16, 2009, Leigh delivered her eulogy for her father.

> Growing up, we could not have asked for a better dad. He was always there to lift us up onto his shoulders for a better view, or give us a piggyback no matter how the walk. He loved to take us apple picking, and encouraged us to climb as high as we could. He always showed us how to reach for that next branch, giving us the courage even when it looked too far away. He was our Santa Clause, all year round. He was the brilliant professor dad with the calculator watch.
>
> Our dad always believed in us. For me, I know I was kind of a rebel at times, but he always looked at the things I did, and would rationalize them as great things that would happen to us. If I did something well, he would inspire me to want to make them even better. Why just 98 percent on the chemistry test, when you know you could get 100 percent. If they went poorly, he showed me how it could not be a greater opportunity for an alternative life plan. He would even go as far as to prove to me that my differences were extraordinary. He would work so hard to convince me that I was unique. He gave me the greatest gift in the world, which is the fire to strive for excellence, and the tools to help me succeed.
>
> Dad had so much patience. If there was something he didn't know or couldn't do, he would work so hard to try to understand it, and grasp what it meant to us. Right after my parents' divorce, it was just me and my dad here in Chicago. I remember the first time he took me shopping for clothes in high school. He really had no idea what he was doing, but he was so invested, and he worked so hard to help me. And when he broke his foot, and couldn't walk to work, I felt so proud to be there to drive him in before school, and then to pick him up to bring him home. And if I got caught up in emotions, he was always so kind and so practical. For instance, if I had misplaced my homework or my wallet, he would take a deep breath, say

it was alright, and then ask me where was the last place I had seen it. My dad had all the patience in the world.

My dad was brilliant. He always made the most interesting analogies and his theories on life were sometimes unanticipated. But, I would think about them for a day or two even, and then something would click, and I'd see his words clearer than water. He was always so right, and he knew the answer well before anyone else. I loved him so much.

He meant so much to everyone around him. Dad was the most selfless man.

Now, I have just begun to understand his place in the world of science. I'm in my fourth year at MIT, and I lived in the same dorm as my dad, and I have the same professors as he had forty years ago. Talking to him last week, on my first day of school, he was excited to hear all about it. I like to think he put me here to continue his legacy, and I will try my hardest because I know it is what he would have wanted. It's going to be difficult to carry on without him, but his enthusiasm and his love will live with us forever. He's given us so much, and we will make him proud.

And even though dad passed away so suddenly—we didn't even get to say good-bye—he prepared us well for difficult times like these. He showed us how to be strong, reach higher, and push through while standing tall. Most of all, he showed us that there's no better thing in life, than having a big heart.

My Journey to Visit Malcolm

I held the crystal ball close to my heart. It was as if I was inside that crystal ball with Malcolm. I was overwhelmed in his presence. He asked me to write down everything he dictated to me. There were tears in his eyes, stricken with grief when he spoke of words in his usual soft speaking voice.

I quickly wrote down words that came from him, the chain, the bioterrorism, the human experiments, biological warfare protocol, deadly *Yersinia pestis*, the vaccine, the POISON and the confession. There was old laughter, new fears; there were betrayals and vengeance of God; there were angels that whispered the truth. In the end, I saw Malcolm carrying a cross on his back, marched in insurmountable pains with all other human sufferers, until he was finally laid down to rest.

I was in tears. I asked for his forgiveness. It was my selfishness that drove us apart, the blindness that made us vulnerable. I did not accompany him through the gates of hell. I did not hold his hands when he was lying there in pain. I did not say good-bye when death came upon him so alone and helpless. I was not there when injustice engulfed his life with a fury.

I will soon start my journey to New Orleans to visit Malcolm. I need to be with him. I need to hold his hands and read aloud the passages from this book to him and to the world, the book that he asked me to write, the book in which he told of his life, his principles, his honor, his integrity, and his love and finally his struggle to preserve human dignity and human rights.

Leigh, Malcolm and Brooke in 2008

Jesus said,

"Then you will know the truth,
And the truth shall set you free."

To Malcolm, my best friend and my love,

This Book Shall Set You Free.

The Truth Shall Set You Free.

AND YOU ARE FINALLY FREE AT LAST

****

Blessed are the poor in spirit, for theirs is the kingdom of heaven.
Blessed are those who mourn, for they will be comforted.
Blessed are the meek, for they will inherit the earth.
Blessed are those who hunger and thirst for righteousness, for they will be filled.
Blessed are the merciful, for they will be shown mercy.
Blessed are the pure in heart, for they will see God.
Blessed are the peacemakers, for they will be called children of God.
Blessed are those who are persecuted because of righteousness, for theirs is the Kingdom of heaven."

Matthew 5:3 Bible

****

RESOURCES & SUGGESTED READING

KaiserHealthnews.org, May 26, 2006

Francis Boyle "Biowarfare and Terrorism", (2005), Clarity Press.

Mark Wheelis, Lajos Rozsa and Malcolm Dando *"Deadly Cultures: Biological Weapons Since 1945"* (2006), Harvard University Press, pp. 284–293, 301–303

Rick Titball and Diane Williamson in "Yersinia pestis (Plague) vaccines." in *Expert Opin. Biol. Ther* (2004), 4:(6): 965-973.

http://www.education.yahoo.com/reference/encyclopedia/entry/streptoc

G. J. Russell-Jones "Oral Vaccine Delivery" (2000) in the Journal of Controlled Release 65 (2000), p49-54.

Appendix 1: Agent Profile Form, 7/25/2007

JUL 25 2007 7/25 P1 627-02

The University of Chicago
Institutional Biosafety Committee
[illegible]
Phone [illegible]
Agent Profile Form
[illegible]

Completion of this form is required for each agent requiring **Biosafety Level 2 (BL2) conditions or higher, or if the biohazard will be administered to animals.** Personnel must be advised of special hazards and must read and follow the required practices and procedures described in this form. This form must be available for laboratory staff and, if applicable, animal care staff..

I. Project

Principal Investigator: Olaf Schneewind, M.D. Ph.D.	**IBC Protocol Number:** 627 **IACUC Protocol Number:** 71065 ☐ N/A

Project Title: Targeting of Yop Proteins by Yersinia enterocolitica

Location where agent will be used:

Laboratories **Building(s):** Carlson, CLSC **Room number(s):** JS22B, 601

Animal Facilities ☐ N/A ☐ Barrier Facility ☐ Non-barrier Facility ☒ Biosafety Facility

II. Agent

1. a. **What is the biohazardous agent?** Y. enterocolitica

   b. **What is the corresponding Risk Group*?** ☐ RG1 ☒ RG2 ☐ RG3

   Please refer to the latest version of the NIH Guidelines for Research Involving Recombinant DNA Molecules at http://www4.od.nih.gov/oba/Rdna.htm .

   **Please note that this institution cannot accommodate RG4 agents.*

2. **CONTAINMENT--Each project within a laboratory must be assigned a biosafety containment level to which the project must adhere. The purpose of containment is to reduce or eliminate exposure of laboratory workers, other persons, and the outside environment to potentially hazardous agents. The most important element of containment is strict adherence to standard microbiological practices and techniques. Persons working with infectious agents or potentially infectious materials must be aware of the techniques required for handling such material safely.**

   a. **Please indicate the laboratory Biosafety Level* appropriate for this biohazardous agent:**

   ☐ BL1 ☐ BL1 w/BL2 practices* ☒ BL2 ☐ BL2 w/BL3 practices* ☐ BL3

   **Please describe the additional BL2 or BL3 practices in Section IV, questions 1 and 4.*

   b. **If the biohazardous agent will be administered to animals, indicate the Animal Biosafety Level*:**

   ☐ N/A ☐ ABSL1 ☒ ABSL2 ☐ ABSL3

   **Refer to the CDC/NIH Biosafety in Microbiological and Biomedical Laboratories (4th edition) http://www.cdc.gov/od/ohs/biosfty/bmbl4/bmbl4toc.htm or the latest version of the NIH Guidelines for Research Involving Recombinant DNA Molecule , http://www4.od.nih.gov/oba/Rdna.htm to determine the biosafety containment level.*

JUL 25 2007 627-02

3. **Will the biohazardous agent be transported between buildings?**

☐ No

☒ Yes: Biohazardous materials must be triple-packaged. *Please see the guidelines for triple-packaging in the CDC/NIH Biosafety in Microbiological and Biomedical Laboratories (4th edition)* *http://www.cdc.gov/od/ohs/biosfty/bmbl4/b4ac.htm*

4. a. **Is this biohazardous agent a recombinant agent?**

☒ No: **Skip to #5**

☐ Yes

b. **In many cases, there may be no known hazards associated with a recombinant agent and/or its products. However, to the best of your knowledge, please describe all potential hazards associated with exposure to the agent and/or its products, considering the expressed transgenes, mutations or other genetic alterations.**

Page 2 of 10

JUL 25 2007 627-02

5. a. **Can the biohazardous agent (wildtype and/or recombinant) infect humans?**

☐ No: **Skip to Section III**

☒ Yes

b. **What disease is caused by this agent?**

Yersiniosis

Reiter Syndrome

c. **What are the symptoms of an infection?**

Yersinia entercolitica can infect the intestines of humans by oral contamination and can cause intestinal lymph adenitis as well as damamge the intestinal epithelium. These lesions cause diahrrhea and/or vomiting as well as fever. Human infections with Y. entercolitica are not frequent;mostly infants and immunocompromised patients are in danger of contracting these infections, particularly those undergoing iron substitution therapies. Retter Syndrome occurs in individuals carrying the HLA B27 allele resulting in uveitis, urethritis as well as arthritis.

d. **How can staff be exposed to the agent** (i.e., ingestion, mucosal contact, inhalation, injection or other percutaneous contact, contact with animal waste, animal bite)?

Ingestion and direct injection into the bloodstream

e. **The following measures are required in the event of an exposure:**

- Notify PI/Supervisor
- Notify UCOM (UC Office of Occupational Medicine, L-156, 702-6757)
- Notify ARC if the exposure is through an animal bite
- Form 45 must be completed (*Form 45 and instructions for completing and submitting the form can be found at http://safety.uchicago.edu/3_1Frameset.html*)

What additional measures should be taken in the event of an exposure to this agent?

f. **Is there anyone who should not enter a room in which this agent is being used** (e.g., individuals who are immunocompromised, pregnant, etc.)?

Indivduals carrying the HLA B27, immunocompromised individuals, indivduals undergoing iron subtitution therapy as well as infants should not enter the laboratoreis where work is being conducted with Y. enterocolitica.

6. **Is pre-exposure vaccination available?**

☒ No

☐ Yes*: **Please specify.**

*Please contact UCOM to make arrangements for vaccination.

JUL 25 2007 627-0

7. **Is pre-exposure serum testing required?**

☒ No

☐ Yes: Please contact UCOM to make arrangements for serum testing. *Please note that this institution cannot accommodate serum banking.*

8. **Will staff be monitored for infections?**

☒ No: Surveillance is not appropriate for this agent

☐ Yes: **Please explain.**

9. a. **Is post-exposure treatment available?**

☐ No: **Skip to Section III**

☒ Yes

b. **What is the treatment?**

Ciprofloxacin

c. **How is the treatment administered?**

500 mg of ciprofloxacin twice daily taken orally for 10 days

d. **What are the potential risks of treatment?**

The major adverse effect seen with use of is gastrointestinal irritation, common with many antibiotics. Studies have not been done in humans. However, use is not recommended during pregnancy since fluoroquinolones have been reported to cause bone development problems in young animals.Some of the fluoroqninolones are known to pass into human breast milk. Since fluoroquinolones have been reported to cause bone development problems in young animals, breast-feeding is not recommended during treatment with these medicines. Use is not recommended for infants or children younger than 18 years of age since fluoroquinolones have been shown to cause bone development problems in young animals.

III. Animal Work

1. **Will the biohazardous agent be administered to animals?**

☐ No: **Skip to Section IV**

☒ Yes: **Species:** **Mice** Number of animals: **960**

JUL 25 2007

627-02

2. **Agent Administration**

a. **Route:**

☐ IM ☐ IN ☒ IP ☐ IV ☐ SQ ☐ Retro-orbital

☒ Other: **intragastric gavage**

b. **Dose (pfu, mg/kg, etc.):** 1 x 10e2 - 1 x 10e9 cfu

c. **Volume:** 0.500 ml

d. **Frequency:** 1

e. **Total # of doses:** 1

f. **Time between doses:** N/A

3. **Potential for transmission of agent**

a. **How is the agent shed (urine, feces, respiratory secretions)?**

Yersinia enterocolitica is shed with the feces.

b. **How long does it persist in the environment?**

It is expected that animals shed the organisms for 7 days.

c. **How long will the animals be considered biohazardous?**

7 days

d. **Can animals treated with these agents be in the same room with untreated animals or animals treated with other agents simultaneously (assuming they are otherwise compatible)?**

☒ Yes

☐ No: **Please explain.**

JUL 25 2007 627-02

4. a. **Will the animals need to be transported between facilities after administration of the agent?**

☒ No **Skip to Section IV**

☐ Yes: The following actions are required:

- The Institutional Animal Care and Use Committee (IACUC) must approve the transport of animals
- Animals must be contained and transported in accordance with Animal Transport Guidelines http://ors/IACUC/policiesProc/AnimalTransportGuidelines.html
- The IBC must approve transport of these animals
- If the animals will be transported in the Medical Center, Infection Control must approve the transport

b. **Please provide scientific justification for the transport of these animals.**

c. **Please describe the method of containment and the route of transport.**

JUL 25 2007 627-02

IV. Safety Practices

1. **PERSONAL PROTECTIVE EQUIPMENT (PPE)/SAFETY EQUIPMENT**

Please check the appropriate biosafety level and, if applicable, animal biosafety level for this agent. PPE requirements are listed under each level. If Section II.3.a. of this form indicates BL1 w/BL2 practices or BL2 w/BL3 practices, please list additional PPE that staff members will be required to use. All staff must comply with these precautions when working with this agent.

	Biosafety Level			Animal Biosafety Level		
Required PPE	☐ BL1	☒ BL2	☐ BL3	☐ ABSL1	☒ ABSL2	☐ ABSL3
Safety Goggles	X	X	X		X	X
Surgical Mask				X	X	
Face Shield*			X			X
Respirator**			X			X
Gloves	X	X	X	X	X	X
Lab Coat	X	X	X			
Shoe Covers			X	X	X	X
Tyvek Sleeves			X	X	X	X
Protective Suit/Gown			X	X	X	X
Handwash Sink***		X	X		X	X
Eyewash	X	X	X	X	X	X
Biosafety Cabinet****			X		X	X
Pull Chain Safety Shower	X	X		X	X	
Shower Out			X			X
Additional PPE						

* Face Shield should be used when working with non-human primates or tissues/cells from the animals.

** OSHA Respiratory Protection and Fit Testing is required of all individuals who must wear a respirator. *Respirator fit testing is available in UCOM at 702-6757, L-156.*

*** Hands-free sinks required for BL3/ABSL3.

**** Required if there is a potential for aerosolization *(including drawing up into syringe).*

PLEASE NOTE: Staff members working with animals must comply with the PPE guidelines of the animal facility in which they are working. The guidelines are posted on the entrances of each facility.

JUL 25 2007 627-0

2. **DECONTAMINATION PROCEDURES**

a. **Which of the following measures are taken during routine decontamination of the work area and pertinent equipment?** *Guidance on decontamination procedures can be found on the Safety and Environmental Affairs website at http://safety.uchicago.edu/2_7Frameset.html* .

- ☐ Treat with 10% bleach* (made fresh daily) for at least 20 minutes
- ☒ 70% ethyl/isopropyl alcohol for 5 minutes
- ☐ Chlorine Dioxide (Clidox)** ☐ Glutaraldehyde (Cidex)** ☐ Phenolic (Lysol)** ☐ Quatricide**
- ☐ UV Light for _____ minutes
- ☒ Autoclaving* at 120 C temperature for 45 minutes
- ☐ Other:

* Please note that materials containing large amounts of bleach should not be autoclaved due to the potential for autoclave damage and generation of toxic by-products.

** Per manufacturer instructions

b. **If the biohazardous agent will be administered to animals, how will the cages be decontaminated?**

- ☐ N/A
- ☐ Standard Rodent Barrier methods (cage washed but not autoclaved)
- ☐ Standard Non-rodent methods (cage washed but not autoclaved)
- ☒ Standard Biosafety Facility methods (cage autoclaved at 120 C temperature for 45 minutes)
- ☐ Other:

JUL 25 2007 627-02

2. **BIOSAFETY CABINET**

A biosafety cabinet is required for the following type of work:

- **All work that has the potential for generating aerosols** *(includes drawing up into syringe)*
- **Work with rodents at ABSL2**
- **All work conducted under BL3/ABSL3 conditions or higher**
- **All work involving human and non-human primate cells/blood/tissues**

Please note that staff should be instructed in the safe operation of the biosafety cabinet and other applicable equipment (centrifuge with sealed rotors, centrifuge cups, etc.) that are used to handle biohazardous agents. *Please note that Appendix A of the BMBL at http://www.cdc.gov/od/ohs/biosfty/bmbl4/bmbl4toc.htm contains information regarding the safe operation of the different classes of BSCs.*

a. **Please indicate the type of biosafety cabinet that will be used for this project.** Contact the Safety and Environmental Affairs Office (702-9999) if you have questions regarding your biosafety cabinet.

☐ I ☒ IIA1 ☐ IIA2 ☐ IIB1 ☐ IIB2 ☐ III

b. **What is the last date that your biosafety cabinet was certified*?** 4/9/2007
*Biosafety cabinets must be certified annually.

VII. General Manipulations (How will the biohazard be used?)

If applicable, please provide supplemental materials, including publications, which might be helpful to the IBC in its review process.

Please provide a brief, lay-language description of the manipulations involved with these biohazardous agents (i.e., cell culturing, cloning, rDNA work, vector production, use in animals). **The description should be sufficient to allow the IBC to perform a risk assessment of this research project. Detailed information such as buffers, significance, etc., is not necessary.**

rDNA manipulations will be done in E. coli under standard laboratory conditions. Once produced, rDNA will be introduced in to Yersinia strains using standard protocol. All Yersinia strains will be routinely grown at 26°C in laboratory medium. Experimental analysis of Yersinia will be done by growing the bacteria at 26 °C followed by a temperature shift to 37°C to express the TTSS and subsequent lysis (fractionation of bacterial cell) of the bacteria to analyze TTSS using standard protocols (ie, western analysis). For animal infections, bacteria will be grown at 26 degrees as stated above and then diluted in Phosphate Buffer Saline prior to intraperotineal injection. Bacterial suspensions will be loaded into 1/2 cc insulin syringes immediately prior to injection in animal room Biosafety cabinet. Once mice are infected, mice will be monitored for illness over a 7 day period.

To study the regulation of villous M cell development, we will use oral gavage with antibiotics (streptomycin) or heat-treated non-pathogenic bacteria (E. coli DH5alpha strain). Chervonsky lab has found that E. coli DH5aplpha induce intestinal M cells when orally administered to mice after exposure to water and one minute of boiling. Oral administration of these reagents (0.5 ml/mouse) is the only way to put them in direct contact with intestinal flora and epithelium. Preliminary experiments in Chervonsky lab [IACUP 73641] found a single dose of 0.5g/kg of streptomycin to be non-toxic for 25g BALB/c mice. To examine the role of induction of additional M cells by stressed commensal flora (E. coli lysates in water) in mucosal immunity, we will challenge mice intra-gastrically with Y. enterocolitica via

JUL 26 2007 627-02

oral gavage. Mice previously gavaged (at 24-36 hrs time point) with 20 ODU/ml E. coli in water, 20mg/mouse of Streptomycin, or PBS control will be infected with up to 10e9 CFU of Y. enterocolitica intragastrically by oral gavage (0.5 ml). Mice will be sacrificed at 24, 48, 72, and 96 hrs and at later time points if needed (day 5, 7, 9, 14) after infections. The number of bacteria in the liver, spleen, and small intestine will be determined. Mice will be observed immediately after oral gavage to be sure that they do not have liquid in their lungs. Mice will be monitored daily after infection. If mice show significant weight loss (>20%), very rough coat, abnormal posture or marked reductions in mobility upon stimulation animals will be sacrificed by the investigator. Moribund mice are immediately sacrificed by CO_2 inhalation and cervical dislocation. All animals that were infected with Y. enterocolitica, irrespective of whether the animals develop signs of acute disease or not, will be euthanized immediately after the 14 day observation period.

JUN 15 2007 627-02

VIII. Staff Group

Please list the name of each staff member. This should include all individuals who will be directly involved in carrying out the work described in the protocol. ALL staff members must read and sign this protocol.

Once the protocol has been approved, any changes in staff should be submitted to the IBC on an amendment form. Note: the Principal Investigator is the only individual who may amend the protocol. If this protocol involves a select agent, please note that all staff members must be screened and approved prior to access to the select agent and other specialty procedures may apply. Contact the IBC for guidance.

READ BEFORE SIGNING

Your signature indicates the following:

- ✓ **You have thoroughly read this protocol submission**
- ✓ **You have sufficient knowledge and are sufficiently trained to perform the responsibilities for which you have been assigned**
- ✓ **If training is required, this will be completed prior to your involvement in this project**
- ✓ **You are aware of and follow the appropriate procedures for the safe handling of the potentially hazardous agents used in this protocol**
- ✓ **You fully understand the steps necessary following any spills or potential exposures with the agents described in this protocol**

Staff Member	Staff Signature
Olaf Schneewind	[signature]
Kelly Riordan	
Alexander Chervonsky	
Kristy Skurauskis	[signature]
Bill Blaylock	B. Blaylock
Nathan Miller	Nathan C. Miller
Lauriane Quenee	QUENEE
Dan Gingold	

JUN 15 2007

IX. Signatures

[illegible]

[illegible]

Principal Investigator:

I understand my responsibility with regard to laboratory safety and certify that the protocol, as approved by the IBC, will be followed during the period covered by this research project. Any future changes will be submitted for IBC review and approval prior to implementation. I understand that this protocol will be reviewed periodically; it is my responsibility to complete and submit in a timely manner the survey form used for the periodic IBC review.

[signature] 6/15/07

Signature of Principal Investigator **Date**

Appendix 2: "Plague at University of Chicago", 10/14/2008

The University of Chicago

Institutional Biosafety Committee
5751 S. Woodlawn Avenue
McGiffert House, Second Floor, MC 1108
(773)-834-5850

Select Agent Institutional Biosafety Committee Certification

Investigator:	Olaf Schneewind
Office Address:	Microbiology CLSC 607
Protocol Number:	**1051**
Protocol Title:	**Plague at University of Chicago**
Approval Date:	**10/07/2008**
Type of Submission:	New
Risk Group:	RG3
Biosafety Level:	BL3
Animal Biosafety Level:	ABSL3

Protocol Status: Approved

The research protocol described above has been reviewed by the SA-IBC with the results as indicated. Any changes to this protocol must be submitted for review to the SA-IBC. SA-IBC protocols are approved for 3 years; however, continued approval for that period is contingent on receipt of PI response to the annual renewal survey.

David Pitrak, M.D.
SA-IBC Chair

OCT 14 2008
Date

I. Study Contact Information:

PI Name: Olaf Schneewind		Unique ID #: 10811
Appointment: Professor, Principal Investigator		Department:
Office Address: 920 E. 58th street		E-Mail: oschnee@bsd.uchicago.edu
Building: CLSC	Room #: 607	Mailcode: 60637

In case of an emergency, contact the PI at:		
Phone: 4-9060	Pager:	Home Phone: (b)(6)

Oversight/Alternate Contact: In the absence of the PI, who will assume responsibility for the ongoing day-to-day oversight and supervision and personnel.

Personal health status may impact an individual's susceptibility to infection, ability to receive immunization, or prophylactic interventions, and this will vary for each biohazardous agent and each person. Therefore, all laboratory personnel should be provided with information regarding diseases or conditions, including pregnancy, that may predispose them to infection or increase their risk of more severe disease in the event an infection occurred with the agent being used. Individuals having these conditions should be encouraged to self-identify to Occupational Medicine or their physician for appropriate guidance and counseling.

Personnel must be advised of special hazards and must read and follow the required practices and procedures described in this form. This form must be available for laboratory staff and, if applicable, animal care staff.

Nature of Submission: ☒ New ☐ Amendment* ☐ Resubmission--Protocol # 1051

**An amendment submission form must accompany any changes made to this form.*

Title of Protocol: Plague at University of Chicago

I. Study Contact Information:

PI Name: Olaf Schneewind		Unique ID #: 10811
Appointment: Professor, Principal Investigator		Department:
Office Address: 920 E. 58th street		E-Mail: oschnee@bsd.uchicago.edu
Building: CLSC	Room #: 607	Mailcode: 60637

In case of an emergency, contact the PI at:		
Phone: 4-9060	Pager:	Home Phone: (b)(6)

Oversight/Alternate Contact: In the absence of the PI, who will assume responsibility for the ongoing day-to-day oversight and supervision and personnel.

Name: Lauriane Quenee	e-Mail: lquenee@bsd.uchicago.edu	
Phone: 4-0566	Pager:	Home Phone: (b)(6)

II. Project:

A. Location of proposed work/experiments:

- ☐ H.T. Ricketts Laboratory
- ☒ Cummings BL3 Laboratories
- ☒ Carlson ABSL3/BL3 Biosafety Facility

B. Project is funded:

☐ Internally
☒ Externally
Awarded? ☒ Yes ☐ No
Grant Agency: NIAID ←
Tracs ID #: 26020
☐ Other, please indicate research sponsor: Novartis (37977)

C. Does this project involve large scale (>10 liters of culture at one time) research or production?

☒ No
☐ Yes: Additional guidelines apply. *Please see Appendix K of the NIH guidelines for Good Large Scale Practices at* http://www4.od.nih.gov/oba/rac/guidelines_02/Appendix_K.htm.
If 'Yes', please indicate the Biosafety Level:
☐ BL1 ☐ BL2 ☐ BL3

D. Does this project involve the acquisition of human tissues directly from patients?

☐ No
☒ Yes, please indicate IRB protocol number: 15672A

E. Research Overview: Please provide a lay-language overview of your research as it pertains to the use of the biohazardous agents.

Yersinia pestis remains a major threat for bioterrorism and biological warfare. Because Americans are not immunized against plague, we rely on basic research for the development of a vaccine, immunotherapies and therapeutics that can be administered to people that have been exposed to plague. At present, there is no vaccine available in the United States.

Our research efforts include identifying new targets for the development of vaccines and therapeutics against *Y. pestis*, the causative agent of plague disease. To this end, we will generate mutant strains of *Y. pestis* (via transposon library or direct mutagenesis) and test the ability of these mutant bacteria to cause disease in animal (*in vivo* assays) or to interact with host cells in tissue culture (*in vitro* assays). Using these *in vitro* assays *Y. pestis* genes will be linked to the different steps of the bacterial life cycle (invasion of host cells, immune system evasion, etc...). Once virulence factors are identified, we will then use these virulence factors as targets to develop vaccines and therapeutic molecules. To do this, proteins involved in bacterial pathogenesis will be purified and used as antigens to stimulate a host immune response to destroy/eliminate *Y. pestis* in/from the host.
Initial screening will be performed by using the attenuated (non virulent) *Y. pestis* KIM D27 (RG2/BSL2). Confirmation of vaccine efficacy and virulence requirement will be performed by using *Y. pestis* CO92 (RG3/BSL3). Previous work has already identified factors that are essential for the pathogenesis of Plague disease. In addition, we have identified 6 potentially surface expose proteins that are encoded in the HPI locus of *Y. pestis*. These proteins may serve as protective antigens. Herein, the role of each of these factors in protective immunity will be assessed purifying the protein and using it vaccine subunit for mice immunization, followed by plague challenge experiments. As there is several type of plague bacteria spread world wide, we will test our vaccine efficacy on different strains of *Yersinia pestis*: Harbin 35 (NR-639), Nepal 516 (NR-640) and ZE94-2122 (NR-635).

Details of the scientific approach:

- **Study of *Y. pestis* virulence factors**
  - ➢ Identification of mutations that abrogate type III secretion of *Y. pestis*.
  - ➢ Creation of tools for measuring *Y. pestis* type III secretion *in vitro and in vivo*.
  - ➢ Measuring *Y. pestis* injection of human blood cells.
  - ➢ Identification of mutations in the HPI that abrogate virulence in animals models of infection.
- **Study of vaccine candidates and immune response for *Y. pestis*.**

- To identify *Y. pestis* surface proteins that serve as protective antigens, we will purify surface proteins and inject them into laboratory animals (mice, rats, guinea pigs, rabbit, NHP (following FDA two animal rule)) to measure their role as protective antigens in a subsequent lethal challenge with virulent *Y. pestis*. We will examine the role(s) of surface proteins in the pathogenesis of plague by infecting animals with mutant yersiniae and measuring the dose required for disease progression.
- To quantify the humoral and cellular immune responses of experimental animals to purified *Y. pestis* surface protein antigens, we will measure antibody production during infection and after vaccination of selected surface antigens and quantify the B- and T-cell response to immunization. The role of humoral and cellular immunity will be determined in passive transfer experiments.
- To investigate the immune mechanisms against *Y. pestis* surface proteins, the role of IL-10, TNF-α, macrophages or adaptive immune cells during plague infection and immunization will be analyzed using mice with specific defects in the immune response.

- **Development of protective index experiments that measure the immune protective ability of sera from vaccinated animals.**
  - We will use mouse protective index experiments (immune sera injected in naïve mice prior challenge with *Y. pestis*) to establish correlations between immune sera and protection against plague in an *in vivo* model. We will test the ability of immune sera to prevent *Y. pestis* from infecting tissue culture cells or cells from whole blood as an *in vitro* way of correlation for protective immunity.
- **Study of therapeutics against plague**
  - We will pursue the development of therapeutic compounds by characterizing their efficacy *in vivo* in mouse models of bubonic and pneumonic plague. In addition, we will analyze whether *Y. pestis* mutants which have spontaneously acquired resistance to these compounds retain virulence in a mouse infection model.

III. Hazardous Agents: Please indicate in the tables below, ALL hazardous agents (regardless of risk level or whether they are Select Agents) that will be utilized in the project. This list should include: bacteria, virus, fungi, parasites, transactive peptides. Also, please submit an Agent Profile Form for each of the agents listed in the three tables. Agent Profile Forms for some agents can be found at http://ors.bsd.uchicago.edu/IBC/index3.html?content=forms/index.html. If the Agent Profile Form is not available for the agent you are using, either modify an existing form (when using an attenuated strain of a virulent agent) or complete the Agent Profile Template if using a new agent.

A) Pathogenic Microorganisms (excluding viral vectors):

Agent Profile forms for certain virulent Select Agents can be found at http://ors.bsd.uchicago.edu/IBC/index3.html?content=forms/index.html. If the Agent Profile form you require is not available, please modify an existing form (if using an attenuated strain) or complete the Agent Profile Template form for a new agent.

Genus/Species	If using the Agent Profile form from our website, is it accurate? (Yes/No)	If 'No', please describe any characterized alterations to the genotype and the resultant phenotype variation.
Yersinia pestis CO92 and its variants	Yes	
Yersinia pestis Harbin 35	Yes	
Yersinia pestis Nepal 516	Yes	
Yersinia pestis ZE94-2122	Yes	
Yersinia pestis KIM D27 and its variants	No	KIM strains are attenuated strains of *Yersinia pestis* (they have lost the pgm locus), thus they are considered an RG2/BSL2 agent
E. coli (K12)	N/A	

B) Other potentially infectious agents (e.g., TAT, viral vectors, poly A, etc):

Please complete Agent Profile Form (HTML link to form) for EACH agent listed below and attach to this submission

Agent Genus/Species or that of source organism	Select Agent? (Yes/No)	Pathogen Host Range	BSL 1, 2, or 3	ABSL 1, 2 or 3	Risk Group 1, 2 or 3

C) Organs, tissues or Cell Cultures:

Cell Line/Tissue	Type (Primary, Established or Immortal)	Source (Animal Species)	Helper Cell Line? (Yes/No)
Whole blood	primary	human, mouse, rat	No
Hela cells	immortal	human	No

Approved by SA-IBC 2/15/2008

THP1 cells	immortal	human	No
CD4+/CD8+	primary	mouse, rat	No
Serum	primary	mouse, rat, NHP, guinea pig, rabbit	No
Purified IgG	primary	mouse, rat, NHP, guinea pig, rabbit	No

IV. **Recombinant DNA and Vectors:**

A. **Does this research involve work with rDNA and/or molecular vectors?**
☐ No
☒ Yes, please indicate/describe in the table below:

Vector	Vector Type (e.g., plasmid, virus, phage, transposon, etc.)	Vector Hosts (e.g., *E. coli*, *Salmonella*, mammalian cells, etc.)	Transgene		
			Transgene or Sequence Designation	Source (species/genus)	Expression Product and/or function (if known)
phage	Mu transposon	E.coli			transposition
pET/pGEX	Expression vector	E. coli			Purification of His tagged or GST tagged proteins
pHSG	Expression vector	Yersinia pestis			Expression of Yersinia protein
pCVD442/pSAC	Non replicative vector	E.coli/Yersinia pestis			Allelic exchange

B. **Will this research involve the deliberate transfer of a drug resistance trait to a pathogenic organism (pathogenic to humans, animals or plants) for which no alternative drugs are available?**
☐ No
☒ Yes, please list the antibiotic resistance markers (be sure to include this information in the appropriate Agent Profile Form): Kanamycin/Ampicillin/Chloramphenicol for KIM (RG2) strains
Kanamycin/Ampicillin for RG3 strains

C. **Will this research involve the generation (other than breeding) of transgenic or knock-out animals?**
☒ No
☐ Yes, please answer question 1-3 below:

1. **Please describe the construct(s) that will be developed in your laboratory and used to generate the transgenic or knock-out animals.** *Be sure to include the rDNA description in the table above in IV A.*

2. **These animals will be generated by:**
   ☐ **PI and/research staff** (please be sure to describe procedures in section)
   ☐ **University of Chicago Transgenic Core Facility**
   ☐ **Outside vendor/collaborator**

3. **Will the transgenic (uninfected) animals and/or excreta pose any risk to individuals having contact with them?**
   ☐ No

☐ Yes, please describe the risk:

V. Viral Work

A. Does this research involve the use of viruses or viral vectors?

☒ No

☐ Yes, answer questions 1-3 below.

1. **Are you using any helper viruses, helper plasmids or viral packaging cell lines? If "YES", please list in Section IV.C.**

   ☐ No

   ☐ Yes, please describe (i.e., ecotropic, amphotropic):

2. **What is the host range for the wild-type and/or recombinant virus?**

3. **Is the virus replication-competent?**

   ☐ Yes

   ☐ No, answer questions 3a and b

   Certain recombinant replication-defective viral vectors (e.g. vectors based on adenovirus and murine retrovirus) can produce recombinant replication-competent virus through homologous recombination with packaging genes or endogenous wild-type virus. For a discussion of this topic see Gene Therapy 10, 706-711(2003) and Gene Therapy 6, 709-712 (1999).

   a) If no, please describe the methods that will be used to test viral preparations for the presence of replication-competent virus. *For further information, please see Testing Requirements for Viral Vectors guidelines at* ***http://ors.bsd.uchicago.edu/IBC/resources/Viral_Vector_Table_Version1.doc*** .

   b) Please indicate the frequency of testing:

VI. Animal Work

A. Will any of the biohazardous agent(s) listed be administered to animals?

☐ No

☒ Yes, indicate the agents below

List of Agents Administered:

Agent	Animal Species/Protocol Number	Route of administration	Dose (pfu, cfu, mg/kg, etc.)	Volume	Frequency	Total # doses	Time between doses
Yersinia pestis	Mouse (71533 and 71808)	IV, SC, IN	10 to 10^8 cfu	100ul, 100ul, 20ul	1	1	N/A
Yersinia pestis	Rat (71560)	ID, IN	10 to 10^8 cfu	100ul, 20ul	1	1	N/A
Yersinia pestis	Guinea pig (71911)	SC, IN	10 to 10^8 cfu	100ul, 100ul,	1	1	N/A

B. Indicate the potential for transmission of each agent listed:

Agent	By what route is agent shed (e.g. urine/feces, respiratory secretions)?	How long does it persist in the environment?	How long will animals be considered biohazardous?
Yersinia pestis	Bubo fluid, CSF, sputum, feces, urine	Sensitive to moist heat or dry heat, sensitive to dryness, survival in blood (100 days, in carcasses up to 270 days)	From injection to autoclave

VII. General Manipulations (How will the biohazard be used?): If applicable, please provide supplemental material, including publications, which might be helpful to the IBC in its review process.

Please provide a brief, <u>lay-language</u> description of the manipulations involved with these biohazardous agents (i.e., cell culturing, cloning, rDNA work, vector production, use in animals). **The description should be sufficient to allow the SA-IBC to perform a risk assessment of this research project. Detailed information such as buffers, significance, etc., is not necessary. For each of the procedures, indicate the type of Biosafety Cabinet that you will be utilizing.**

Y. pestis RG3 will be manipulated under BSL3 containment in a class IIb2 BSC. *Y. pestis* RG2 (attenuated strains) will be manipulated under BSL2 containment. All animal work will be performed in a Class II Biosafety Cabinet under BL2 or BL3 containment depending on the RG of the strain.

Yersinia pestis CO92 and its variants: Select Agent, RG3, BSL3
Yersinia pestis Harbin 35: Select Agent, RG3, BSL3
Yersinia pestis Nepal 516: Select Agent, RG3, BSL3
Yersinia pestis ZE94-2122: Select Agent, RG3, BSL3
Yersinia pestis KIM D27 and its variants: NON-Select Agent, RG2, BSL2

As KIM D27 is an attenuated strain, it is not the best candidate strain to test either virulence or vaccine efficacy. That strain will not be used in animals at high doses. Maximum dose for KIM D27 injection into animals will be 10^6 CFU.

- **Genetic manipulation of *Y. pestis***

Please note that for all *Y. pestis* RG3, antibiotic resistance markers used for genetic manipulation will be Kanamycin or Ampicillin. Chloramphenicol will not be used in RG3 but will be used in RG2 *Yersinia* (attenuated strains).

➢ Generation of transposon libraries.

Bacteriopaghe or non-replicative plasmids are used to introduce transposable genetic elements in *Yersinia*. Those transposable elements are then inserted into the chromosome or endogenous plasmids of the bacteria. Insertion of transposable elements, depending on the site of insertion, will disrupt the function of the inserted gene. Such transposable elements are designed to transpose and insert themselves once into the bacterial genome. They cannot transpose away from the initial transposition site, allowing the mutation to be stable over time. Bacteria carrying the insertion are selected using the resistance marker specific for the transposable element (Kanamycin; Chloremphenicol, Ampicillin) and screened for differential phenotype.

➢ Directed mutagenesis

▪ Lambda Red

The Lambda Red system allows performing the disruption of targeted genes. The parental strains of *Yersinia* are transformed with a plasmid expressing a transposase. A PCR product is generated by amplifying a resistance cassette (Kanamycin or Ampicillin) flanked by regions that are homologous to the to-be disrupted gene. After electroporation in the bacteria and upon transposase activity, the PCR product is inserted in the bacterial chromosome at the site where homologous recombination is allowed by the flaking regions. Mutated strains are selected using the resistance marker.

▪ Non-replicative vector

A non-replicative vector is design to carry the mutated version of a target gene (deletion, point mutation, genetic tags). This plasmid carries a resistance gene (Ampicillin or chloremphenicol) as well as a counter-selection gene: *sacB*. Homologous recombination between the parental allele and the mutated version is selected upon resistance selection, resulting in the full integration of the vector into the bacterial chromosome. The resolution of this integration is performed by counter selection of the co-integrated strains on culture media containing 5% sucrose. Upon the *sacB* gene activity, sucrose is toxic for the bacteria. In order to grow, the *sacB* gene will be excised from the chromosome by homologus recombination between the parental allele and the mutated version. As a result, these 2 recombination steps allow the replacement of the parental allele by the mutated one. The final mutant strain does not carry any resistance marker.

➢ Complementation

Replicative vectors are use to complement *Yersinia* strains. These plasmids carry the parental version of a target gene as well as a resistance marker for plasmid retention (Ampicillin, Kanamycin, chloramphenicol). They are electroporated into *Yersinia* in order to complement the expression of a mutated gene.

➢ Reporter protein

Replicative vectors might carry a modified version of a protein (fused with GFP, His tag, FlAsH tag) as well as resistance marker for plasmid retention (Ampicillin, Kanamycin, chloramphenicol). They are electroporated into *Yersinia* in order to allow the expression of a tagged protein by the bacteria.

- ***In vitro* assays**

➢ Type III Secretion assay

In order to measure type III secretion *in vitro*, bacteria are grown in culture media at 37°C in absence of calcium for 3 hours. Under these conditions, the type III proteins are secreted into the culture media. Culture supernatant and bacterial pellet are then collected and submitted to TCA protein precipitation. Proteins are visualized by SDS-PAGE electrophoresis (denaturing conditions) using Coomassie blue staining, silver staining, or specific immunoblotting methods.

➢ Tissue culture cell assay

In order to measure type III injection *in vitro*, bacteria are incubated with tissue culture cells (HeLa cells) at 37°C for 3 hours. The cells are then fractionated using a specific detergent (Digitonin 1%). Proteins that have been injected from the bacteria into the cells are visualized in the cytosol of the injected cell by immunoblotting detection methods and by microscopy.

In order to measure cell invasion, monolayers of cell lines are infected with *Y. pestis* and placed in the 37°C incubator for 15 min to 3 hours. Uptake is measured by treating the infected cells with gentamicin (an antibiotic that cannot permeate the cell line). External bacterial cells are killed and the cell culture monolayers are washed

Approved by SA-IBC 2/15/2008

with sterile PBS and lysed with detergent. Bacteria phagocytosed by the eukaryotic cells survive this treatment and can be subsequently plated on agar plates. Enumeration of surviving (phagocytosed) bacteria compared to initial inoculum will be used to determine %phagocytosis.

➢ *Yersinia pestis* phagocytosis assay in whole blood

We will draw fresh whole blood from laboratory animals (mice, rats) or human volunteers (IRB #15672A). Blood will be aliquoted into screw-cap micro tubes (2ml) with a 1ml sterile pipette for the assay. Anticoagulated blood will be incubated with *Y. pestis* at 37°C for 30 min to 5 h. The reaction mixture (*Y. pestis*, antibodies and blood) will incubate on a rotating contraption designed to securely hold micro tubes and resting on an adequate layer of absorbent material. Samples will be taken at select time points with sterile pipette tips and deposited into their appropriate wells of a 96 well, round bottom plate or eppendorf tubes. Up to 8-ten-fold dilutions will be made in 1xHBSS and dilutions plated (manually via single or multi-channel pipettor) on blood agar plates. Colonies will be counted manually. Enumeration of surviving bacteria compared to the initial inoculum will be used to determine % phagocytosis.

- **Protein expression and purification in *E. coli***

Yersinia genes of interest are cloned into *E. coli* expression vector (His-tag, GST-tag). *Yersinia* proteins are then expressed and purified out of *E. coli* using an affinity column (Nickel for His-tagged proteins or Glutathion for GST-tagged proteins). All of these proteins will be produced in the following manner: amplification of the gene from *Y. pestis* genome using PCR then insertion of the gene into an expression vector, either a pET or pGEX plasmid, then transformation of the *E. coli* strain BL21 (DE3) with the newly constructed plasmid. Protein expression will be induced by the addition IPTG and purified using affinity chromatography, e.g., Ni-NTA beads or Glutathione sepharose. Purified proteins are then adsorbed to adjuvant for use as vaccine formulation in animal experiments.

- **Use of *Y. pestis* in laboratory animals.**

Y. pestis will be inoculated to laboratory animals to evaluate the requirement of virulence factors for plague pathogenesis as well as for testing the prophylactic efficacy of new vaccine strategies or new vaccine candidates and for testing the efficacy of new therapeutics. Inoculation routes and doses vary depending on the animal model used.

➢ Mouse model of plague
  - Mice will be infected intravenously to mimic septicemic plague (dose: from 10 to 10^8 cfu, volume 100ul; location: retro-orbital sinus; manipulation: mice are anesthetized for the procedure, bacteria are injected with 28g needle).
  - Mice will be infected subcutaneously to mimic bubonic plague (dose: from 10 to 10^8 cfu, volume 100ul; location: inguinal fold; manipulation: mice are hand restrained, bacteria are injected with 28g needle).
  - Mice will be infected intranasally to mimic pneumonic plague (dose: from 10 to 10^8 cfu, volume 20ul; location: nostril; manipulation: mice are anesthetized for the procedure, bacteria are pipetted into nostril).

➢ Rat model of plague
  - Rats will be infected intradermally to mimic bubonic plague (dose: from 10 to 10^8 cfu, volume 100ul; location: dorsal lumbar region; manipulation: rats are hand restrained for the procedure, bacteria are injected with 28g needle).
  - Rats will be infected intranesally to mimic pneumonic plague (dose: from 10 to 10^8 cfu, volume 20ul; location: nostril; manipulation: rats are anesthetized for the procedure, bacteria are pipetted into nostril).

➢ Guinea pig model of plague
  - Guinea pigs will be infected subcutaneously to mimic bubonic plague (dose: from 10 to 10^8 cfu, volume 100ul; location: hind leg; manipulation: guinea pigs are hand restrained for this procedure, bacteria are injected with 28g needle).
  - Guinea pigs will be infected intranasally to mimic pneumonic plague (dose: from 10 to 10^8 cfu, volume 100ul; location: nostril; manipulation: guinea pigs are anesthetized for this procedure, bacteria are pipetted into nostril).

➢ Post-infection procedures in mice, rats, and guinea pigs
  - Clinical observation:

Infected animals will be observed twice daily for 14 days for the development of disease symptoms and time-to-disease will be recorded. Monitoring frequency will increase as symptoms start to appear. All animals that will be infected with *Yersinia pestis*, irrespective of whether the animals develop symptoms or not, will be euthanized by

CO2 after 14 days.

- Post-mortem procedures

Depending on the studies conducted, animals will be subjected to pathological examination. At different time points during the experiments, animals will be euthanized by CO2.

- Bacterial dissemination

To measure the bacterial dissemination and replication in the infected animals, blood, liver, spleen and lungs will be removed, double bagged and homogenized. Serial dilutions of homogenized organs will then be plated in order to allow bacterial enumeration.

- Histopathology

To visualize the gross and microscopic histopathology of infected animals, liver, spleen and lungs will be removed and fixed in 10% formalin solution. Formalin solution will be refreshed once and after 2 weeks of fixation, samples of the fixed tissues will be plated to determine that no bacteria had survived the formalin treetment. Sterile tissues will be processed for histopathology.

➢ Animal number

Animal numbers for a 3 year period, has been estimated as follow:
Mice: 8860 and 1680; guinea pigs: 2187; rats: 1784. Details on the use of these animals are available in the IACUC protocols (71533, 71808, 71911 and 71560, respectively).

VIII. Staff Group: Please list all individuals who will be directly involved in the work described in this protocol.
Note: *This protocol is restricted to work in locations which require extensive safety training prior to access being granted. In some cases, annual re-training is required to maintain access. This training includes, but is not limited to: proper ingress/egress procedures, PPE, proper use of biosafety cabinet, spill procedures, decontamination procedures, and specific health monitoring procedures. For a complete list of training that will have to be completed and SOP's that need to be reviewed prior to access being granted, please go to URL.*

Name	**Position** (Post-doc, technician, student)
Olaf Schneewind	Professor PI
Lauriane Quenee	Research Project Manager
Nancy Ciletti	Senior Research Technician
Derek Elli	Senior Research Technician
Bryan Berube	Junior Research Technician
Melanie Marketon	Visiting professor
Timothy Hermanas	Junior Research Technician
Nathan Miller (RG2/BSL2)	Graduate Student
Bill Blaylock (RG2/BSL2)	Postdoctoral fellow
Will DePaolo	Research associate
Maloolm Casadaban (RG2/BSL2)	Research associate
Dominique Missiakas	PI
Stefan Richter	Research Project Manager
Helene Louvel	Postdoctoral fellow
Anthony Mitchell	Junior Research Technician
Jen Stencel (RG2/BSL2)	Graduate Student

IX. PI Signature:

The undersigned investigator is responsible for:

- providing adequate training and supervision of staff in microbiological techniques and practices required to ensure safety and for procedures in dealing with accidents,

- enforcing federal regulations regarding laboratory safety for all persons who work under his/her direction,
- correcting work errors and conditions that may result in the release of rDNA materials or infectious agents and ensuring the integrity of the physical containment,
- reporting to the IBC, any adverse event, such as a work related injury or exposure,
- ensuring that co-investigators, if any, employ the necessary safeguards to protect laboratory personnel, students, and the community from potential hazards posed by the project,
- and, the investigator must ensure that all staff members have read this protocol.

A copy of the protocol and each Agent Profile Form must be available to all staff members. For further information regarding physical safety issues, laboratory safety, emergency response and training, please contact the Office of Safety and Environmental Affairs (702-9999) or the Biosafety Officer for the University of Chicago (834-7496).

Principal Investigator:

I understand my responsibility with regard to laboratory safety and certify that the protocol, as approved by the SA-IBC, will be followed during the period covered by this research project. Any future changes will be submitted for SA-IBC review and approval prior to implementation. I understand that this protocol will be reviewed periodically; it is my responsibility to complete and submit in a timely manner the survey form used for the periodic SA-IBC review.

Signature of Principal Investigator **Date**

Appendix 3: Pathogenic Profile of KIM D27

OCT 07 2008 1051

The University of Chicago

Select Agent Institutional Biosafety Committee

Agent Profile Form

SECTION I - INFECTIOUS AGENT

NAME: *Yersinia pestis* **KIM D27 strain (RG2) and its variants, attenuated strains.**

SYNONYM OR CROSS REFERENCE: Plague, Peste, Bubonic plague

CHARACTERISTICS: Gram negative rod-ovoid 0.5-0.8 µm in width and 1-3 µm in length, bipolar staining (safety pin appearance), facultative intracellular, non-motile

SECTION II - HEALTH HAZARD

PATHOGENICITY: Zoonotic disease; bubonic plague with lymphadenitis in nodes receiving drainage from site of flea bite, occuring in lymph nodes and inguinal areas, fever, 50% case fatality if untreated; may progress to septicemic plague with dissemination by blood to meninges; secondary pneumonic plague with pneumonia, mediastinitis, and pleural effusion; untreated pneumonic and septicemic are fatal

EPIDEMIOLOGY: Wild rodent plague in North America, South America, Africa, Near and Middle East, Central and Southeast Asia, Indonesia; plague foci in USSR; urban plague controlled in most areas; human plague occurred recently in Africa; endemic in Burma and Vietnam; sporadic cases in North and South America following exposure to wild rodents or their fleas (no human-to-human transmission in USA since 1925)

HOST RANGE: Humans, > 200 mammalian species

INFECTIOUS DOSE: Unknown; depends upon route of infection

MODE OF TRANSMISSION: Result of human intrusion into zoonotic (sylvatic) cycle or by entry of rodents or infected fleas into human's habitat and bite of infected fleas; domestic pets can carry plague-infected fleas; contact of commensal rodents and their fleas with sylvatic rodents may result in epizootic and epidemic plague; handling of infected tissues; airborne droplets from humans or pets with plague pneumonia; careless manipulation of laboratory cultures; person-to-person transmission by human fleas; percutaneous injection via contaminated SHARPs.

INCUBATION PERIOD: From 2 to 6 days; may be a few days longer in vaccinated individuals; for primary plague pneumonia, 1 to 6 days, usually short

COMMUNICABILITY: Fleas may remain infective for months; bubonic plague not usually transmitted directly from person-to-person; pneumonic plague may be highly communicable under appropriate climatic conditions (overcrowding facilitates transmission)

SECTION III - DISSEMINATION

RESERVOIR: Wild rodents (rats) are the natural reservoir; lagomorphs (rabbits, hares) and carnivores may be a source of infection to humans

ZOONOSIS: Yes - bites of fleas from an infected animal; contact or being bitten by an infected animal

VECTORS: Wild rodent fleas, especially the oriental rat fleas (*Xenopsylla cheopis*); occasionally by human fleas (*Pulex irritans*)

SECTION IV - VIABILITY

DRUG SUSCEPTIBILITY: Sensitive to streptomycin, tetracycline, chloramphenicol (for cases of plague meningitis), kanamycin (for neonates)

DRUG RESISTANCE: Generally not a concern; a multi-drug resistant strain (MDR) mediated by transferrable plasmid has been isolated

SUSCEPTIBILITY TO DISINFECTANTS: Susceptible to many disinfectants - 1% sodium hypochlorite, 70% ethanol, 2% glutaraldehyde, iodines, phenolics, formaldehyde

PHYSICAL INACTIVATION: Sensitive to moist heat (121° C for at least 15 min) and dry heat (160-170° C for at least 1 hour)

SURVIVAL OUTSIDE HOST: Blood - 100 days; human bodies - up to 270 days

SECTION V - MEDICAL

SURVEILLANCE: Monitor for symptoms; presumptive diagnosis by visualizing bipolar staining, ovoid, gram-negative organisms in sputum or material aspirated from bubo; FA and ELISA test; PHA using Fraction-1 antigen

FIRST AID/TREATMENT: Antibiotic therapy in early stages (8 to 24 hours after onset of pneumonic plague); secondary infection or suppurative bubo may require incision and drainage

IMMUNIZATION: Although field trials have not been conducted to determine the efficacy of licensed vaccines, experience has been favourable; immunization is recommended for personnel working regularly with culture of /*Y. pestis*/ or infected rodents, boosters are required every 6 months if high risk continues; protection against pneumonic form is limited

PROPHYLAXIS: Chemoprophylaxis using tetracyclines or sulfonamides; for close contacts of pneumonic cases

All patients will receive empiric therapy for any febrile illness until the diagnostic evaluation reveals an alternative diagnosis and/or symptoms resolve. Fever should be used as a signs that someone might be infected, as should a suppurative lymphadenitis, or a febrile respiratory illness. Admission is not necessary for patients with fever alone. Tetracycline or doxycycline for 10 days is adequate.

Patients should be admitted for nausea/vomiting with dehydration, other signs of sepsis, or pulmonary symptoms as they require respiratory isolation. For patients who a seriously ill requiring hospitalization or are unable to take oral medication, therapy will be IM streptomycin 15 mg/kg q 12 hours for 10 days or until another diagnosis is made. Although SM is usually given IM, IV is a safe alternative. Although SM is considered the most active aminoglycoside, gentamicin is also effective and is more frequently given as an IV agent.

For patients who are allergic to streptomycin, have other relative contraindications, or for whom an oral medication is strongly preferred, tetracycline (500 mg PO qid) or doxycycline (100 mg PO bid) can be given as an alternative.

In the rare patient with meningitis, the initial regimen should include chloramphenicol (1 g IV q 6 hours). Other drugs with in vitro activity include fluoroquinolones and trimethoprim/sulfamethoxazole.

Patients with pneumonia, sepsis, and meningitis may require the addition of other empiric antibiotics to cover for common pathogens not associated with laboratory exposure.

Post-exposure Prophylaxis, Asymptomatic Patient - If a lab worker has an inoculation injury or a significant splash exposure, doxycycline 100 mg PO bid should be administered for 10 days. If there is an allergy to tetracyclines, cipro 500 mg PO bid can be given.

ANTIBIOTIC RESISTANCE MARKERS: Kanamycin, Ampicillin, Chloramphenicaol

SECTION VI - LABORATORY HAZARDS

LABORATORY-ACQUIRED INFECTIONS: none

SOURCES/SPECIMENS: Bubo fluid, blood, sputum, CSF, feces, urine

PRIMARY HAZARDS: Direct contact with cultures and infectious materials from humans or rodents; infectious aerosols or droplets generated during manipulation of cultures and infected tissues and in the necropsy of rodents; accidental auto-inoculation; ingestion

SPECIAL HAZARDS: Bites by infected fleas collected from rodents

SECTION VII - RECOMMENDED PRECAUTIONS

CONTAINMENT REQUIREMENTS: *Y. pestis* strain KIM D27 and its variants carry a deletion of 102 kb pgm locus and therefore display a severe defect in virulence for all animals, including humans (Skrzypek & Straley 1995 J. Bacteriol. 177:2530). On 3/14/03, *Y. pestis* strain KIM D27 was made exempt from CDC registration and is considered a RG2 strain.
Biosafety level 2 practices, containment equipment and facilites for all activities involving the handling of potentially infectious clinical materials and cultures

PROTECTIVE CLOTHING: Gloves should be worn when handling field-collected or infected laboratory rodents and when there is the likelihood of direct skin contact with infectious materials.

OTHER PRECAUTIONS: Special care should be taken to avoid the generation of aerosols during the necropsy of animals; necropsy should be conducted in a biological safety cabinet; insecticide treatment when collecting animals (living or dead) for testing

OCT 07 2008 1051

SECTION VIII - HANDLING INFORMATION

SPILLS: Allow aerosols to settle; wearing protective clothing, gently cover spill with paper towels and apply 1% sodium hypochlorite, starting at perimeter and working towards the centre; allow sufficient contact time (30 min) before clean up. See Spill Protocol SOP for additional details.

DISPOSAL: Decontaminate before disposal; steam sterilization, chemical disinfection of contaminated liquid (freshly prepared 1% bleach, f.c.)

STORAGE: In sealed containers that are appropriately labeled

SECTION IX – MISCELLANEOUS INFORMATION

Date Originally Approved:

Approved by: **University** of Chicago Select Agent Institutional Biosafety Committee

---

If Agent Profile Form has been modified for an attenuated strain

Date Approved: 10.07.08

Approved by: University of Chicago Select Agent Institutional Biosafety Committee

Appendix 4: Staff Addition to Vaccination Protocol, 8/21/08

The University of Chicago

Institutional Biosafety Committee
5751 S. Woodlawn Avenue
McGiffert House, Room 214, MC 1108
(773)-834-5850

Institutional Biosafety Committee Certification

Approval Date: 08/21/08 ← 10/6/08 sent to NIH-OBI

Investigator: Olaf Schneewind
Office Address: Microbiology
CLSC 1117B

Amendment Number: **AD 04**
Protocol Number: **627**
Protocol Title: Targeting of Yop Proteins by *Yersinia enterocolitica*

Risk Group: RG2
Biosafety Level: BL2
Animal Biosafety Level: ABSL2

Nature of Amendment: **Staff Addition: Staff Additions: Staff Additions: Kristy Skurauskis, Bryan Berube, Claire Cornelius, Nancy Ciletti, Timothy Hermanas, Malcom Casadaban & Laura Satkamp**

Amendment Status: Approved

The amendment to the research protocol described above has been reviewed by the IBC with the results as indicated. Any additional changes to this protocol must be submitted to the IBC and approved prior to initiation. Thank you for your cooperation.

Lenora Hallahan (signature) — AUG 2 2008 — 8/20

Lenora Hallahan
Administrator for Regulatory Compliance

Date

AUG-20-2008 17:30 From: 7733340655 To:773 834 0659 P.1/3

AUG 2 . 2008 627-04

The University of Chicago

Institutional Biosafety Committee

[illegible] S. Woodlawn Avenue

[illegible]

[illegible]

Amendment Submission Form

For Changes in Protocols Involving the Use of Biohazardous Materials in Research

Please submit this form, **typed** and completed in full, to McGiffert House, 2nd Floor, 5751 S. Woodlawn Avenue. **Amendments to approved protocols may not be initiated until IBC approval has been obtained.** The IBC reserves the right to determine whether proposed changes are substantive, and to request further information or a new protocol submission, as appropriate.

Principal Investigator: Olaf Schneewind	**Department:** Microbiology
Protocol title: Targeting of Yop Proteins by Yersinia enterocolitica	**Protocol #:** 627

I. Proposed Changes (Please check all that apply)

☐ **Change in PI to:** | **Reason for change:**

In addition to this form (signed by the original PI), **please submit the following sections of the Protocol Submission Form:**

- **Section I** with revised contact information *(new PI and, if applicable, alternate contact)*
- **Section VIII** with revised staff list with signatures
- **Section IX** that has been read and signed by the new PI

☒ **Staff Changes**

In addition to this form, **please submit Section VIII of the Protocol Submission Form** that has been read and signed by the new staff members.

Staff being added:	**Staff being removed** (if applicable):
Kristy Skurauskis	
Bryan Berube	
Claire Cornelius	
Nancy Ciletti	
Timothy Hermanas	
Malcolm Casadaban ←	
Laura Satkamp	

☐ **Location of work:** *Please include revised protocol sections where room numbers are referenced.*

Location being added:	**Location being removed** (if applicable):
Bldg(s):	**Bldg(s):**
Room(s):	**Room(s):**

☐ **Other** *(change in vectors, hosts, DNA, containment, risk group, etc.)*

Please explain the proposed changes and provide sufficient rationale for each change. As the Committee may not be familiar with all vectors, nature of the DNA, hosts, etc., please provide sufficient identification, nature and background to allow for an adequate review. In addition to this form, **please submit revised Protocol Submission Form and/or Agent Profile Form** pages to incorporate these changes. ***Please Note: If the proposed change involves a change in the Risk Group and/or Biosafety Level, the protocol MUST BE submitted on the new forms. Please contact the IBC Office for further information.***

AUG-20-2008 17:30 From: 7738340655 To:773 834 0659 P.3/3

AUG 2? 2008 627-04

VIII. Staff Group

Please list the name of each staff member. This should include all individuals who will be directly involved in carrying out the work described in the protocol. **ALL staff members must read and sign this protocol.**

Once the protocol has been approved, any changes in staff should be submitted to the IBC on an amendment form. Note: The Principal Investigator is the only individual who may amend the protocol. If this protocol involves a select agent please note that all staff members MUST be screened and approved prior to access to the select agent and other special procedures may apply. Contact the IBC for guidance.

<u>READ BEFORE SIGNING</u>

Your signature indicates the following:

- ✓ **You have thoroughly read this protocol submission**
- ✓ **You have sufficient knowledge and are sufficiently trained to perform the responsibilities for which you have been assigned**
- ✓ **If training is required, this will be completed prior to your involvement in this project**
- ✓ **You are aware of and follow the appropriate procedures for the safe handling of the potentially hazardous agents used in this protocol**
- ✓ **You fully understand the steps necessary following any spills or potential exposures with the agents described in this protocol**

Staff Member	Staff Signature
Kristy Skurauskis	
Bryan Berube	
Claire Cornelius	
Nancy Ciletti	
Timothy Hermanas	
Malcolm Casadaban ←	
Laura Satkamp	

AUG-20-2008 17:30 From: 7738340655 To:773 834 0659 P.2/3

AUG 21 2008 627-04

11. Signature of PI: The Principal Investigator is responsible for informing staff members of the above changes to the protocol and, if applicable, training staff on new procedures related to the above changes. [illegible] original signature of the Principal Investigator.	
[signature]	Date: 8/20/08

Appendix 5: The Autopsy Report

THE UNIVERSITY OF CHICAGO MEDICAL CENTER

Division of Anatomic Pathology
Room E603 - MC6101
5841 South Maryland Avenue
Chicago, IL 60637-1470
Phone: (773) 702-6327 Fax: (773) 702-9903

Autopsy Report

Patient Name:	CASADABAN, MALCOLM			Accession #:	A09-132
Med. Rec. #:	3101880	Client::	1 - University of Ch	Expiration Date:	9/13/2009 17:06
DOB(Age):	8/12/1949 (Age: 60)	Location:	BIER (UCH)	Autopsy Date:	9/15/2009
Gender:	M	Service:	Emergency Medicine	Completed:	10/23/2009
Physician:	JENNIFER CHAN				

Post Mortem Hours: 41
Authorized By: Brooke Casadaban
Relationship to Patient: Daughter
Autopsy Restrictions: None
Prosector: Ilyssa Okrent Gordon, M.D. Ph.D.

Final Anatomic Diagnosis

I. Genetic hemochromatosis
  A. Markedly increased iron stores in non-cirrhotic liver.
  B. Markedly elevated serum ferritin, iron, total iron binding capacity, and percent iron saturation.
  C. Homozygous for C282Y mutation of HFE gene by genetic testing.
  D. No evidence of iron deposition in any other organ.

II. Clinical history of leukocytosis with left shift and sepsis
  A. Antemortem blood cultures positive for an attenuated Yersinia pestis strain (KIM D27), and nutritionally variant streptococci.
  B. Area of mucosal prolapse of the sigmoid colon, associated with a small area of possible mucosal damage or ulceration and focal bacterial organisms (Gram positive).
  C. Enlarged, soft spleen (522 grams).
  D. Post mortem blood cultures positive for Klebsiella pneumonia.
  E. Gallbladder unremarkable with patent bile duct.
    1. Bile stained gastric and duodenal contents and mucosa (diffuse) and patchy distal esophageal bile stained mucosa.
    2. Grey-green mucosal discoloration of cecum, right colon, left colon, and sigmoid mucosa (sparing of transverse colon mucosa) without evidence of mesenteric thrombi.
  F. Left and sigmoid colon with diverticular disease and serosal adhesions to left lower quadrant peritoneum.
  G. Appendix present and unremarkable.

III. Clinical history of respiratory failure
  A. Lungs (right: 990 grams; left: 980 grams) with congestion and edema but no pneumonia or hemorrhage.
  B. No evidence of pulmonary embolism.

IV. Clinical history of congestive heart failure and hypertension

A. Cardiomegaly (601 grams) with mild biventricular dilatation and petechial hemorrhage in the myocardium.
B. Left anterior descending coronary artery with distal eccentric atherosclerotic plaque and 60% stenosis.
C. Kidneys (right: 237 grams; left: 212 grams) with bilateral nephrosclerosis, cortical and cortico-medullary petechial hemorrhage, and focal sinus hemorrhage.
D. Patent foramen ovale (0.8 cm)

V. Clinical history of abnormal bladder ultrasound
A. Bladder with superficial mucosal hemorrhage at dome.
B. Prostate enlarged with nodular hypertrophy
C. Distal prostatic parenchymal hemorrhage with tracking into posterior aspect soft tissue.

VI. Multinodular thyroid (37 grams)

Peter Pytel, MD
Attending Pathologist
Electronically Signed Out by Peter Pytel, MD

pp

Procedures/Addenda

Addendum **Date Ordered:** 11/25/2009
Date Complete: 11/25/2009

Addendum Diagnosis

THE FOLLOWING OUTSIDE CONSULTATON REPORT WAS RECEIVED FROM MAYO CLINIC, MAYO MEDICAL LABORATORIES DATED 11/25/09:

PATIENT NAME	PATIENT NUMBER	AGE	SEX	ACCESSION #
CASADABAN, MALCOLM	3101880	60	M	W3286220
ORDERING PHYSICIAN	CLIENT ORDER #			ACCOUNT #
PYTEL, PETER	A09-132			F1011435 and C7011435

COLLECTION	RECEIVED	REPORT PRINTED	SPECIMEN INFORMATION
11/25/09 10:19 A DATE TIME	11/21/09 07:52 A DATE TIME	11/25/09 10:20 A DATE TIME	DATE OF BIRTH: 8/12/1949
Univ of Chicago Hospitals Attn: Pathology Report Office 5841 S. Maryland Ave Room E602 MC 6101 773-702-6327			

TEST REQUESTED	HI LO			REF RANGE	PERFORM SITE*
Iron, Liver Ts					
Iron, Liver Ts	H	14672	ug/g dry wt	200-2400	SDL
Hepatic Iron Index	H	4.4	umol/g/yr	<1.0	SDL

Results of Hepatic Iron Index between 1.0-1.9 suggest mild, nonspecific iron accumulation as may be seen in alcoholic liver disease or heterozygous hemochromatosis. Results >1.9 indicate homozygous hemochromatosis or transfusion-related iron overload. Chronic blood loss or frequent phlebotomy will decrease the hepatic iron index.

CASADABAN, MALCOLM **Autopsy Report** A09-132

*PERFORMING SITE

MCR	Mayo Clinic Dpt or Lab Med & Pathology 200 First Street SW, Rochester, MN 55905	Lab Director: Franklin R. Cockerill, III, M.D.

PATIENT NAME	ORDER STATUS	COLLECTION DATE AND TIME
CASADABAN, MALCOLM	Final	11/25/09 10:19 A

ORIGINAL CONSULTATION REPORT ON FILE IN SURGICAL PATHOLOGY LABORATORY.

Addendum **Date Ordered:** 12/28/2009
Date Complete: 12/28/2009

Addendum Diagnosis
THE FOLLOWING OUTSIDE CONSULTATON REPORT WAS RECEIVED FROM CENTERS FOR DISEASE CONTROL AND PREVENTION, ATLANTA, GA DATED 10/23/09:

Diagnosis:
Systemic infection of Yersinia pestis with positive special stain results and immunohistochemical evidence. (see comments)

Addendum Comment

Comments:
The diffuse presence of microthrombi and numerous bacteria in vascular lumens of all organs are highly suggestive of an overwhelming systemic bacterial infection with septic shock. These bacteria are characterized as gram-negative bacilli with Lillie-Twort tissue Gram stain. Warthin-Starry silver stain also highlights these bacilli in the tissue sections examined.

Immunohistochemical (IHC) test using a multi-step indirect immunoalkaline phosphatase technique was performed on various tissues. The primary antibody used in the test was a mouse monoclonal anti-Y. pestis antibody*. Appropriate positive and negative controls were run in parallel. Abundant immunostaining of Yersinia pestis, predominantly in the blood vessels, was seen in all the tissues submitted for testing. Based on the histopathologic features and localization of bacterial antigens, the cause of death can be attributed to a septicemic plague. Correlation with clinical history, epidemiologic information, and other laboratory assays is recommended.

*Immunohistochemical detection of Yersinia pestis in formalin-fixed, paraffin-embedded tissue. Guarner J, Shieh WJ, Greer PW, Gabastou JM, Chu M, Hayes E, Noltke KB, Zaki SR. Am J Clin Pathol. 2002 Feb;117(2):205-9.

THE ORIGINAL CONSULTATION REPORT ON FILE IN THE SURGICAL PATHOLOGY LABORATORY.

Autopsy Summary

This is a 60 year old male with history of diabetes, hypertension, and hyperlipidemia, who presented to the University of Chicago Emergency Department on the morning of 9/13/09 with a four to seven day history of worsening shortness of breath and associated dry cough, fevers, chills, and weakness, as well as orthopnea, lower extremity edema, and abdominal swelling. His condition continuously deteriorated over an approximately 12 hour period when he died with respiratory failure and hypotension progressing to pulseless electrical activity. Laboratory studies showed blood cultures were positive for Gram negative rods and Gram positive cocci in chains in two specimens, as well as yeast in one specimen. The Gram negative rods were later identified as an attenuated Yersinia pestis strain (KIM D27), and the Gram positive cocci were later identified as nutritionally variant streptococci. There was also lactic acidosis, and peripheral blood smear showed high numbers of extracellular bacteria and a leukocytosis of 76.2 K/uL with a marked left shift (22% bands).

At autopsy, the most significant finding was that of markedly increased iron deposition in the liver parenchyma. Post mortem testing on antemortem blood confirmed markedly elevated serum ferritin, iron, total iron binding capacity, and percent iron saturation (serum ferritin 392.530 ng/ml, total iron binding capacity 648mcg/dl, iron 541 mcg/dl, saturation 83.5%). Post mortem genetic testing confirmed the presence of a homozygous C282Y mutation in the HFE gene. All of these findings are consistent with a diagnosis of genetic hemochromatosis. This diagnosis was not clinically apparent during the patient's life.

Numerous organs, including heart, lungs, spleen, and liver, had vascular congestion with evidence of leukemoid reaction and intravascular bacterial organisms, consistent with the antemortem peripheral blood smear showing extracellular bacteria. Although post mortem testing of blood and lung only reveal Klebsiella pneumonia (significance uncertain), further testing on the ante mortem blood culture organisms initially identified as Gram negative rods, were later determined to be of an attenuated Yersinia pestis (KIM D27), after extensive laboratory testing. The portal of entry or source of infection by this organism is unclear. There were no skin lesions and there was no morphologic evidence of pneumonia. No larger ulcerated lesions were found in the oropharynx or gastrointestinal tract. Based on theses findings, the overall history and possibly the presence of focally ulcerated mucosal prolapse in the colon, the portal of entry through the gastrointestinal tract may have to be considered.

The autopsy did not reveal any other findings besides leukemoid reaction with the possibility of sludging and bacteremia to explain the patient's clinical presentation. In particular there are no changes of pneumonia, pulmonary embolic disease or myocardial infarction. Regarding the bladder, the superficial mucosal hemorrhage at the bladder dome, as well as the prostatic parenchymal and soft tissue

THE ORIGINAL CONSULTATION REPORT ON FILE IN THE SURGICAL PATHOLOGY LABORATORY.

Autopsy Summary

This is a 60 year old male with history of diabetes, hypertension, and hyperlipidemia, who presented to the University of Chicago Emergency Department on the morning of 9/13/09 with a four to seven day history of worsening shortness of breath and associated dry cough, fevers, chills, and weakness, as well as orthopnea, lower extremity edema, and abdominal swelling. His condition continuously deteriorated over an approximately 12 hour period when he died with respiratory failure and hypotension progressing to pulseless electrical activity. Laboratory studies showed blood cultures were positive for Gram negative rods and Gram positive cocci in chains in two specimens, as well as yeast in one specimen. The Gram negative rods were later identified as an attenuated Yersinia pestis strain (KIM D27), and the Gram positive cocci were later identified as nutritionally variant streptococci. There was also lactic acidosis, and peripheral blood smear showed high numbers of extracellular bacteria and a leukocytosis of 76.2 K/uL with a marked left shift (22% bands).

At autopsy, the most significant finding was that of markedly increased iron deposition in the liver parenchyma. Post mortem testing on antemortem blood confirmed markedly elevated serum ferritin, iron, total iron binding capacity, and percent iron saturation (serum ferritin 392.530 ng/ml, total iron binding capacity 648mcg/dl, iron 541 mcg/dl, saturation 83.5%). Post mortem genetic testing confirmed the presence of a homozygous C282Y mutation in the HFE gene. All of these findings are consistent with a diagnosis of genetic hemochromatosis. This diagnosis was not clinically apparent during the patient's life.

Numerous organs, including heart, lungs, spleen, and liver, had vascular congestion with evidence of leukemoid reaction and intravascular bacterial organisms, consistent with the antemortem peripheral blood smear showing extracellular bacteria. Although post mortem testing of blood and lung only reveal Klebsiella pneumonia (significance uncertain), further testing on the ante mortem blood culture organisms initially identified as Gram negative rods, were later determined to be of an attenuated Yersinia pestis (KIM D27), after extensive laboratory testing. The portal of entry or source of infection by this organism is unclear. There were no skin lesions and there was no morphologic evidence of pneumonia. No larger ulcerated lesions were found in the oropharynx or gastrointestinal tract. Based on theses findings, the overall history and possibly the presence of focally ulcerated mucosal prolapse in the colon, the portal of entry through the gastrointestinal tract may have to be considered.

The autopsy did not reveal any other findings besides leukemoid reaction with the possibility of sludging and bacteremia to explain the patient's clinical presentation. In particular there are no changes of pneumonia, pulmonary embolic disease or myocardial infarction. Regarding the bladder, the superficial mucosal hemorrhage at the bladder dome, as well as the prostatic parenchymal and soft tissue hemorrhage were likely a result of manipulation and instrumentation in the area. No bladder tumor and no ulceration of the urothelial lining are found.

The patient's diabetes may have been a general risk factor for bacterial infection. In this particular case, though, the presence of hemochromatosis appears crucial in explaining the patient's disease. Yersinia species are known to be siderophores, and thrive in iron-rich environments. There are several reports of patients with hemochromatosis experiencing otherwise unexplained episodes of severe sepsis and infection by other Yersinia species. The patient's undiagnosed hemochromatosis likely provided the environment for an infection with a Yersinia species even though this particular strain of Yersinia pestis is otherwise regarded as attenuated. Overwhelming cytokine release associated with the Gram negative Yersinia sepsis and with the severe leukemoid reaction likely led to the patient's demise.

References:

1: Reinicke V, Korner B. Fulminant septicemia caused by Yersinia enterocolitica. Scand J Infect Dis. 1977;9(3):249-51.
2: Capron JP, Capron-Chivrac D, Tossou H, Delamarre J, Eb F. Spontaneous Yersinia enterocolitica peritonitis in idiopathic hemochromatosis. Gastroenterology. 1984 Dec;87(6):1372-5.
3: Abbott M, Galloway A, Cunningham JL. Haemochromatosis presenting with a double Yersinia infection. J Infect. 1986 Sep;13(2):143-5.
4: Vadillo M, Corbella X, Pac V, Fernandez-Viladrich P, Pujol R. Multiple liver abscesses due to Yersinia enterocolitica discloses primary hemochromatosis: three cases reports and review. Clin Infect Dis. 1994 Jun;18(6):938-41.
5: Collazos J, Guerra E, Fernández A, Mayo J, Martínez E. Miliary liver abscesses and skin infection due to Yersinia enterocolitica in a patient with unsuspected hemochromatosis. Clin Infect Dis. 1995 Jul;21(1):223-4.
6: Weinberg ED. Microbial pathogens with impaired ability to acquire host iron. Biometals. 2000 Mar;13(1):85-9.

Clinical History

This is a 60 year old male with history of diabetes, hypertension, and hyperlipidemia, who presented to the University of Chicago Emergency Department on the morning of 9/13/09 with a four to seven day history of worsening shortness of breath and associated dry cough, fevers, chills, and weakness, as well as orthopnea, lower extremity edema, and abdominal swelling. On presentation, he was afebrile, normotensive and saturating at >95% on nasal cannula, however, he continued to have increased work of breathing and was placed on face mask and eventually intubated after several hours. A chest x-ray obtained at presentation was negative for pleural effusions or pulmonary opacities and showed only a borderline enlarged heart. A later chest x-ray showed moderate cardiomegaly with bilateral pulmonary edema, consistent with congestive heart failure. Brain natriuretic peptide levels were markedly elevated and rising. Blood cultures were positive for Gram negative rods and Gram positive cocci in chains in two specimens, as well as yeast in one specimen. The Gram negative rods were later identified as an attenuated Yersinia pestis strain (KIM D27), and the Gram positive cocci were later identified as nutritionally variant streptococci. There was also lactic acidosis, and peripheral blood smear showed high numbers of extracellular bacteria and a leukocytosis of 76.2 K/uL with a marked left shift (22% bands). The patient continued to have respiratory failure and became hypotensive with decreased saturation and eventual pulseless electrical activity, from which he did not recover after 15 minutes of resuscitative measures. Time of death was 5:06PM on 9/13/09, which was approximately 12 hours after presentation.

Additional clinical information became available after the patient's death and after the autopsy. The patient worked with an attenuated strain of Yersinia pestis (KIM D27) in a research laboratory laboratory.

Gross Anatomical Status
Gross and Microscopic Anatomical Status

External Appearance: The body (weight 265 lbs, height 69 in) is that of a Caucasian male appearing the indicated age of 60 years. There is moderate rigor and dorsal lividity. The scalp exhibits short white hair. The pupils are equal, round, and 0.5 cm in diameter. The sclerae are white. The frontal dentition is unremarkable, but the mouth is not further opened due to rigor mortis. The neck is unremarkable. The chest is symmetrical without increase in the A-P diameter. The abdomen is protuberant. There is mild pitting edema of the lower extremities.

Tubes and external markings: The skin is devoid of scars or lesions. There are two small purpuric lesions over the right abdomen, consistent with bruises of unclear etiology. An endotracheal tube and an orogastric tube are in place with green-brown bilious fluid staining around the mouth. There is a single lumen intravenous catheter in the left antecubital area and a triple lumen catheter in the right groin.

Peritoneal Cavity: The abdominal subcutaneous fat is 3 cm thick. Omental and mesenteric fat is present and there are minimal adhesions between the omentum and the abdominal organs. The abdominal organs occupy their usual positions. The liver edge is at the costal margin. The spleen is at the costal margin. Ascites is absent. There are minimal adhesions between the descending colon and the left lower quadrant peritoneal surface; the peritoneal surfaces are otherwise smooth and glistening.

Pericardial cavity: The pericardial sac contains no fluid. The pericardial surfaces are smooth with focal petechial hemorrhage on the anterior surface. The mediastinum is midline.

Heart and blood vessels: The heart (601 grams) has a moderate amount of epicardial fat and focal petechial hemorrhage on the anterior aspect. The heart is serially sectioned from the apex to 1 cm below the ventricular valves and reveals red-brown, firm myocardium with diffuse congestion throughout the left ventricular myocardium, as well as some mottling in the posterior wall of the left ventricle that focally shows some of the features of an old infarct (<0.2 cm). The left ventricular wall measures 1.8 cm and the right ventricular wall measures 0.5 cm. The heart is then opened along the course of blood flow to reveal a patent foramen ovale with a diameter of 0.8 cm. A gross photograph is taken. The aortic and pulmonic valves are thin and translucent. The tricuspid and mitral valve leaflets are translucent with thin chordae and are devoid of vegetations. The circumference of the tricuspid, pulmonic, mitral, and aortic valves are 15 cm, 10 cm, 12 cm, and 8.5 cm, respectively. The coronary artery orifices are patent. The posterior ventricular circulation is supplied by the right coronary artery, while the left circumflex artery courses just lateral to the left anterior descending artery. There is 60% stenosis of the distal left anterior descending coronary artery. The left circumflex and right coronary arteries are patent. The great vessels arise normally from the base of the heart and are distributed in the usual manner. The ostia of the celiac, superior mesenteric, and renal arteries are patent. The intimal surface of the aorta is smooth and glistening with minimal plaques and calcification in the distal aspect.

Microscopic:

Left ventricular endomyocardium and posterior aspect scars (A15, A18, A19, A27):

- Focal mild subendocardial interstitial scarring with adjacent myocyte hypertrophy.
- Focal moderate atherosclerotic stenosis of a subendocardial arteriole, with 75% occlusion.
- Capillary congestion with leukemoid reaction and rare bacterial rods and cocci.
- No evidence of myocardial iron deposition (iron stain negative)

Respiratory tract: The pleural surfaces of the right lung (990 grams) and left lung (980 grams) are smooth and glistening without adhesions. There is 25 cc serous fluid in the right pleural cavity and 10cc serous fluid in the left pleural cavity. The trachea is opened to reveal pink-red glistening mucosa without ulceration. The larynx and vocal cords have a slight green-grey discoloration and are without edema, ulceration, or obstruction. The bronchial mucosa is pink and glistening. The lungs are serially sectioned to reveal red-brown, crepitant parenchyma, without consolidation, emphysematous spaces, or masses. The pulmonary arteries show no evidence of thromboembolic disease.

Microscopic:

Left lung (A10, A11, A29, A32):

- Capillary and arteriole congestion with leukemoid reaction and numerous intraluminal bacterial rods and cocci.
- Increased septal perivascular macrophages
- Mild focal pulmonary edema in peripheral lower lobe
- Mild emphysematous change.

Right lung (A12-A14, A33-A35):

- Vascular congestion with leukemoid reaction and intraluminal bacterial rods and cocci.
- Increased septal perivascular macrophages.
- Microscopic focus of hemosiderin-laden macrophages in central lower lobe.

Alimentary Tract: The esophageal mucosa is tan, glistening, and lined by longitudinally striated gray mucosa. In the distal third of the esophagus, there are patches of bile stained mucosa. No ulcerations are seen grossly in the esophagus. The gastric mucosa is green-grey with flattened rugal folds and focal mild petechiae. The wall is thin and pliable without masses or lesions. There is bilious content in the stomach and proximal duodenum. The small bowel is opened to reveal transversely folded mucosa. No ulcerations are seen grossly in the stomach. The duodenal mucosa is green-grey, while the distal small bowel mucosa is pink-tan. There is a 0.2 cm white mucosal polyp in the duodenum. The colon contains light-brown, semi-firm stool. The colonic mucosa is notable for diffuse green-grey discoloration extending from the ileocecal valve to the start of the transverse colon, and into the descending and sigmoid colon. The transverse colon mucosa is pink-tan. There are two small mouthed diverticuli filled with soft tan stool in the sigmoid colon. There are also two small polyps in the sigmoid colon, measuring 0.2 and 0.3 cm. Gross photographs of the colon are taken. No ulcerations are seen grossly in the bowel. The mesentery was serial sections to show normal blood vessels without significant stenosis or thrombosis. No significant lymphadenopathy is noted along the GI-tract.

Microscopic:

Distal esophagus (A20):

- Autolyzed esophageal wall with focal parenchymal and intravascular bile accumulation.

Duodenal and colonic polyps (A16):

- Autolyzed mucosa with leukemoid reaction and rare intravascular bacteria.
- Brunner gland hyperplasia and mucosal bile accumulation of duodenum.
- Area of mucosal prolapse of the sigmoid colon. This area is associated with a small area of possible mucosal damage or ulceration and with focal bacterial organisms (Gram positive on Gram stain).

Liver: The liver weighs 2018 grams and has a sharp anterior margin. The external capsule is mahogany brown, smooth and glistening. The liver is serially sectioned to reveal firm brown parenchyma with normal lobular architecture. The intrahepatic vessels and portal vein are patent.

Microscopic:

Liver (A7, A26, A36-A40):

- Autolyzed liver parenchyma with markedly increased parenchymal iron stores, not involving the bile ducts and highlighted by iron stain.
- Multifocal microscopic parenchymal and intravascular accumulations of bacteria (no staining on Gram stain).
- Scattered mutlifocal intravascular and parenchymal calcifications (GMS negative for fungal organisms)
- Intravascular leukemoid reaction.

Biliary system: The gallbladder is filled with 4 cc of green-brown, viscous bile. The wall is thin and pliable with a velvety, green-brown mucosa. There are no calculi. The common bile duct is patent and measures 0.2 cm in circumference.

Pancreas: The pancreas weighs 130 grams and is yellow-brown and firm. The pancreas is serially sectioned to reveal a lobular, yellow-brown parenchyma and sparse fat.

Microscopic:

Pancreas (A8, A41-A43):

- Partially autolyzed pancreatic parenchyma with no evidence of increased iron stores; iron stain negative.

Urinary tract: The kidneys (right: 237 grams; left: 212 grams) have capsules that strip with ease to reveal multifocal cortical scarring. The kidneys are bisected at the hilum to reveal a cortico-medullary architecture without cysts or masses. Diffuse congestion of the cortex and cortico-medullary junction is noted bilaterally. The urinary bladder is opened to reveal a patch of mucosal hemorrhage at the bladder dome, which does not extend beyond the mucosa. The mucosa is otherwise tan, glistening, and trabeculated, without ulceration, masses or other lesions.

Microscopic:

Left and right kidney (A4, A5, A24):

- Mild arterionephrosclerosis and interstitial edema.
- No evidence of iron deposition on iron stain

Bladder (A23):

- Recent submucosal hemorrhage.

Reproductive organs: The prostate is moderately enlarged and has multiple bulging yellow-tan parenchymal nodules adjacent to the prostatic urethra, which is patent and measures 0.2 cm in diameter. The distal aspect of the prostate is involved by diffuse parenchymal hemorrhage, which tracks into the posterior prostatic soft tissue.

Microscopic:

Prostate (A21, A22):

- Acute parenchymal hemorrhage.

Spleen: The spleen weighs 522 grams and has a smooth gray capsule. On cut section, the parenchyma is dark red and semi-firm without lesions.

Microscopic:

Spleen (A6, A25, A28):

- Focal microscopic areas of early necrosis with fibrin, neutrophils and macrophages.
- Partly intracellular appearing degenerating organisms consistent with bacteria (do not stain by Gram staining).

Bone Marrow: The rib, vertebral and sternal marrow is trabeculated, red, and coarsely granular.

Microscopic:

Bone marrow (A1):

- Normocellular bone marrow with trilineage hematopoiesis with myeloid and erythroid left shift to immaturity.
- No bacteria identified by Gram stain.

Endocrine glands: The thyroid weighs 37 grams and is multinodular, glistening and tan-brown. The thyroid is serially sectioned to reveal multiple nodules, some of which are partially calcified. The adrenal glands (right: 12.5 grams; left: 8 grams) have uniform yellow cortices. The adrenals are serially sectioned to reveal a well-demarcated grey medulla.

Microscopic:

Left adrenal (A2) and right adrenal (A3):

- Autolyzed adrenal parenchyma with no evidence of storage iron.

Thyroid (A9):

- Multinodular thyroid with focal calcification

Musculoskeletal: No deformities of the extremities or vertebral column are seen. There is a patch of soft tissue hemorrhage in the chest plate overlying the heart corresponding to resuscitative efforts.

Gross and Microscopic Brain Status

Brain: The extracted brain weighs 1426 gm. The vertebral and basilar arteries show a mild S-shaped deformity and focal atherosclerotic plaque deposition. Otherwise the main arterial vessels of the circle of Willis are intact and unremarkable. There is no evidence of any significant stenotic lesions or thrombosis. The meninges are delicate and unremarkable. The surface of the brain shows no areas of softening, focal atrophy, or mass effect. No midline shift or herniation is found. The cerebral hemispheres are serially sectioned into coronal slices, the brain stem and cerebellum into transaxial slices.

These sections show normal brain parenchyma with normal anatomy of the gray and white matter structures. No focal lesions are seen. No atrophic changes or mass effect are noted on the cut sections. The spinal cord is grossly unremarkable.

Microscopic:

Pituitary (A17):
- Without diagnostic abnormality
- No evidence of iron deposition by iron stain

Cerebellum (B1):
- Without diagnostic abnormality

Medulla (B2):
- Without diagnostic abnormality

Spinal cord (B3):
- Without diagnostic abnormality

Hippocampus (B4):
- Without diagnostic abnormality

rgr

Slide Key

Part: A

1	Bone marrow
2	Left adrenal
3	Right adrenal
4	Left kidney
5	Right kidney
6	Spleen
7	Liver
8	Pancreas
9	Thyroid and parathyroid
10	Lung, left upper lobe
11	Lung, left lower lobe
12	Lung, right upper lobe
13	Lung, right middle lobe
14	Lung, right lower lobe
15	Left ventricle
16	Duodenal and colonic polyps
17	Pituitary
18	Left ventricle
19	Left ventricle
20	Distal esophagus
21	Distal prostate
22	Proximal prostate
23	Bladder wall
24	Kidney
25	Spleen
26	Liver
27	Left ventricle
28	Spleen
29	Left lung
30	Lymph nodes
31	Adipose tissue and lymph nodes
32	Left lung
33	Right lung
34	Right lung
35	Right lung
36	Liver

37 Liver
38 Liver
39 Liver
40 Liver
41 Pancreas
42 Pancreas
43 Pancreas

Part: B
1 Cerebellum
2 Medulla
3 Spinal Cord
4 Hippocampus
5
6
7

Special Studies, Reports, and Consultations
Saved serum from pre mortem studies was used for additional post mortem clinical chemistry testing. These studies showed:

Serum ferritin 392.530 ng/ml (normal 20-300)
Total iron binding capacity 648mcg/dl (normal 230-430)
Iron 541 mcg/dl (normal 40-160)
% Saturation 83.5 (normal 14-50)

Post mortem genetic testing confirmed the presence of a homozygous C282Y mutation of the HFE gene.

Appendix 6: Mandatory Vaccination Program at the Chicago National Lab

The University of Chicago

Institutional Biosafety Committee

5841 S. Maryland Avenue
AMB S-152 ♦ MC 1108
Chicago, IL 60637

Minutes of September 30, 2004 Meeting
1:00 PM in [redacted]

In attendance:

Voting Members

Kenneth Thompson
Richard Hiipakka
Mark Abe
Malcolm Casadaban
George Daskal
Clara Gartner
Jean Greenberg
Tong-Chuan He
Helena Mauceri
Mary Ellen Sheridan
Gopal Thinakaran
Craig Wardrip

Ex-Officio Members

Steve Beaudoin
Russell Herron
David Pitrak
Markus Schaufele
Steve Seps

Guest

Debra Anderson
Kristen DeBord
Judd Johnson
Olaf Schneewind

Staff

Pamela Postlethwait
Bill Pugh
Jennifer Swanson

Absent:

Voting Members

James Mastrianni
Louis Philipson

Ex-Officio Members

Michael Holzhueter

Staff

None

I. **Presentation by Dr. Olaf Schneewind**
Dr. Schneewind provided an overview of his research and the BSL3 facilites.

II. **Protocol Review:**

Both of the protocols reviewed at this meeting involve the use of select agents and will be conducted in BSL3 facilities under similar SOPs. The Committee discussed the following issues, which pertain to both of these protocols.

The Committee discussed whole facility decontamination. In previous discussions, it was proposed that Vaporous Hydrogen Peroxide (VHP) decontamination would be performed annually and between the use of different select agents. However, as routine decontamination and environmental monitoring will be conducted, VHP decontamination will only be performed in response to off-normal conditions, spill of the select agent or detection of contamination through environmental monitoring. There are no regulations that require annual or between agent VHP decontamination.

The Committee discussed environmental monitoring procedures for the facilities. Quarterly monitoring will be conducted by the University Safety Office. When conducting environmental monitoring, the Safety Office follows departmental SOPs which outline sampling locations and number of samples to be collected. This information, however, was not included in the investigator's protocol submissions. Several members suggested that reference to these SOPs should be provided. After discussion of this issue, the Committee felt that summaries of the monitoring reports would be sufficient.

A member questioned whether or not environmental monitoring would be conducted after VHP decontamination. As biological indicators will be used during the decontamination process, environmental monitoring is not necessary.

The Committee discussed the need for decontamination of the ductwork. According to the Commissioning Agent, a plan should be in place for decontaminating the ductwork beyond the HEPA filter. In response to this issue, the University Safety Office indicated that contamination of the ductwork would be a rare event and that a plan would be developed at the time of need. Several members felt that a written plan should be incorporated into the protocol, as recommended by the Commissioning Agent. After discussing this issue further, the Committee agreed that an SOP for decontamination of the ductwork should be developed and included in the protocol prior to approval of the research.

The SOP for Fever Watch was discussed by the Committee. A member felt that the Fever Watch consent form should emphasize the need to seek medical attention in cases of known exposures. The Committee felt that this was not necessary, as this consent was developed strictly for Fever Watch and not for specific exposures. All staff members will be provided training on exposures and whom to contact in the event of an exposure. Furthermore, the two-person rule will be implemented at all times, and the exposed person would be assisted in seeking medical attention.

A member noted an area of the protocol submission form in which the investigator provided reference to an SOP, rather than providing an actual summary of the information. The member questioned whether or not this was appropriate. As the particular SOP is quite detailed, the Committee felt that it was acceptable to simply provide the reference.

The Committee discussed the use of human cell lines. HELA cells will be used, which are considered non-infectious.

The Committee noted items from the Commissioning Report work that need to be completed prior to approval of the research. This includes: testing of the emergency response system; installment and testing of bubble dampers; testing of the autoclave; completion of response to Item #34.

Administrative issues that need to be addressed include: clarification of routine decontamination procedures; removal of references to animal work.

In order to evaluate the SOPs and environmental monitoring, the Committee requested that the investigator submit summaries of two quarterly monitoring reports to the IBC.
The following protocols were reviewed at the meeting, with the disposition noted:

PR# **Category/Investigator/*__Disposition__***

867 **New/Schneewind/*__Pending-Conditions & Stipulations__* __(11-0-0)__**

This research involves the use of *Yersina pestis* strains CO92 and KIM. The goal of this research is to identify new targets for vaccine and immunotherapies by screening for protective antigens and by elucidating the pathogenic features of immune protection.

__Pending Conditions:__
The PI must ensure that the following issues regarding the document entitled "[redacted] are addressed.

1. The U of C response to Item #1 indicates a 10/18/04 target completion date for work related to the second bubble tight damper. Confirmation of installation, testing and proper operation of the damper is needed.

2. The U of C response to Item #5 indicates a 10/18/04 target completion date for testing of the alarm notification time. Confirmation of the testing and proper operation of the emergency response plan is needed.

3. In Item #17, [redacted] assessment notes that a start-up report is needed for the autoclave. Confirmation of testing and proper operation of the autoclave is needed.

4. The U of C Response for Item #34 needs to be completed.

5. In Item #35, [redacted] recommends the development of procedures for decontamination of ductwork. The U of C response notes that contamination of ductwork would be a rare event, therefore, decontamination procedures for the ductwork will be generated upon a report of contamination. The Committee, however, requested that a plan be in place prior to approval of the research. The Investigator must provide a copy of the SOP that describes procedures for decontamination of ductwork. **(Vote 6-5-0)**

__Stipulations:__

The Investigator needs to submit summaries of the first two quarterly monitoring reports.

868 **New/Schneewind/*__Pending-Conditions & Stipulations__* __(13-0)__**

This research involves the use of Bacillus anthracis AMES. The goal of this research is to characterize (1) the role of sortase genes in surface protein anchoring and (2) the role of sortases and surface proteins in invasion of host cells.

The Committee discussed vaccination of staff members working with *Bacillus anthracis*. The University has concerns about mandatory vaccination. However, if the philosophy of the lab is to require vaccination, employees that wish to work with the agent would have to be vaccinated. If an employee refused vaccination, the University's legal office should be consulted, and alternative responsibilities would need to be assigned to the employee.

Pending Conditions:

The PI must ensure that the following issues regarding the document entitled "[redacted] are addressed.

1. The U of C response to Item #1 indicates a 10/18/04 target completion date for work related to the second bubble tight damper. Confirmation of installation, testing and proper operation of the damper is needed.

2. The U of C response to Item #5 indicates a 10/18/04 target completion date for testing of the alarm notification time. Confirmation of the testing and proper operation of the emergency response plan is needed.

3. In Item #17, [redacted] assessment notes that a start-up report is needed for the autoclave. Confirmation of testing and proper operation of the autoclave is needed.

4. The U of C Response for Item #34 needs to be completed.

5. In Item #35, [redacted] recommends the development of procedures for decontamination of ductwork. The U of C response notes that contamination of ductwork would be a rare event, therefore, decontamination procedures for the ductwork will be generated upon a report of contamination. The Committee, however, requested that a plan be in place prior to approval of the research. The Investigator must provide a copy of the SOP that describes procedures for decontamination of ductwork.

Stipulations:

The Investigator needs to submit summaries of the first two quarterly monitoring reports.

ABOUT THE AUTHOR

Joany Chou, Ph.D.

Dr. Joany Chou holds a Ph.D. in virology from the University of Chicago in 1986 and a B.S. from the University of California at Berkeley. Dr. Chou has published numerous scholarly articles on antibiotic resistance transposable elements and later on the biology of human infectious agent, Herpes Simplex Virus I. Dr. Chou also holds patents for several breakthrough discoveries in the areas of gene therapy, tumor therapy and therapy for prevention and treatment of viral diseases.

Dr. Chou was married to renowned scientist, Dr. Malcolm Casadaban, with whom she has two daughters. Dr. Chou fought for and won a major victory for the rights of laboratory researchers in the landmark legal case *Chou v University of Chicago* (2002).

In 2009, Dr. Casadaban, a professor and researcher at the University of Chicago, died tragically from a plague infection in the blood under mysterious circumstances. Dr. Chou has dedicated the years since 2009 to uncovering and exposing the truth behind his death. This book entails her path and endeavor in her investigation in seeking the truth and bringing justice for Malcolm.

27754550R00164

Printed in Great Britain
by Amazon